Synthesis Lectures on Engineering, Science, and Technology

The focus of this series is general topics, and applications about, and for, engineers and scientists on a wide array of applications, methods and advances. Most titles cover subjects such as professional development, education, and study skills, as well as basic introductory undergraduate material and other topics appropriate for a broader and less technical audience.

Emre Tokgoz

Six Sigma for Continuous Improvement in Cybersecurity

A Guide for Students and Professionals

Emre Tokgoz [id]
Farmingdale State College
New York State University
Long Island, NY, USA

ISSN 2690-0300   ISSN 2690-0327 (electronic)
Synthesis Lectures on Engineering, Science, and Technology
ISBN 978-3-031-91029-6   ISBN 978-3-031-91030-2 (eBook)
https://doi.org/10.1007/978-3-031-91030-2

© The Editor(s) (if applicable) and The Author(s), under exclusive license to Springer Nature Switzerland AG 2026

This work is subject to copyright. All rights are solely and exclusively licensed by the Publisher, whether the whole or part of the material is concerned, specifically the rights of translation, reprinting, reuse of illustrations, recitation, broadcasting, reproduction on microfilms or in any other physical way, and transmission or information storage and retrieval, electronic adaptation, computer software, or by similar or dissimilar methodology now known or hereafter developed.
The use of general descriptive names, registered names, trademarks, service marks, etc. in this publication does not imply, even in the absence of a specific statement, that such names are exempt from the relevant protective laws and regulations and therefore free for general use.
The publisher, the authors and the editors are safe to assume that the advice and information in this book are believed to be true and accurate at the date of publication. Neither the publisher nor the authors or the editors give a warranty, expressed or implied, with respect to the material contained herein or for any errors or omissions that may have been made. The publisher remains neutral with regard to jurisdictional claims in published maps and institutional affiliations.

This Springer imprint is published by the registered company Springer Nature Switzerland AG
The registered company address is: Gewerbestrasse 11, 6330 Cham, Switzerland

If disposing of this product, please recycle the paper.

Preface

Events in the cybersecurity world require cybersecurity professionals to implement continuous improvements to existing cyber-defense mechanisms against cyber-attacks while trying to reduce existing costs. Six Sigma is a developed technique that helps to understand existing operations better and implement continuous improvements for cost-effectiveness and improved operations. Defining a problem upon a thorough understanding of the problem, collecting data that relates to the solution of the problem, analyzing the collected data, investigating improvement opportunities, and adopting changes to increase productivity and defense strategies are essential for operational excellence in cybersecurity management. To the best of author's knowledge, this is the first textbook where Lean Six Sigma techniques and Cybersecurity applications meet with the goal of training cybersecurity professionals. Define, Measure, Analyze, Improve, and Control (DMAIC) steps are outlined in this book with examples given in cybersecurity. Noting that six sigma training requires statistics and technology knowledge for data analysis, this textbook covers the associated statistics, six sigma, lean, and technology-related concepts that a cybersecurity student or professional can use to an extent. Waste categories for Lean Six Sigma application purposes are defined for the first time in this textbook. This textbook can be appropriate for a lower or upper-level undergraduate course that depends on the statistics and technology requirements of the course. This book can also be used for six sigma certificate training of six sigma professionals and students.

Readers will be finding examples of six sigma tool applications in cybersecurity that relate to inventory coverage, response time to close and incident analysis, and risk assessment analysis. There are other ideas related to Six Sigma projects and applications provided throughout this book. I hope you will enjoy the reading and examples.

Dr. Emre Tokgoz

Acknowledgement Special thanks to Jennifer L. Bayuk who graciously agreed to share the data used for the six sigma applications that the author applied the DMAIC approach.

Competing Interests The author has no competing interests to declare that are relevant to the content of this manuscript.

About This Book

Events in the cybersecurity world require cybersecurity professionals to implement continuous improvements on existing defense mechanisms against cyber-attacks while trying to reduce existing costs. Six Sigma is a developed technique that helps to understand existing operations better and implement continuous improvements for cost-effectiveness and improved operations. Defining a problem upon a thorough understanding of the problem, collecting data that relates to the solution of the problem, analyzing the collected data, investigating improvement opportunities, and adopting changes to increase productivity and defense strategies are essential for operational excellence in cybersecurity management. To the best of author's knowledge, this is the first textbook where Lean Six Sigma techniques and Cybersecurity applications meet with the goal of training cybersecurity professionals. Define, Measure, Analyze, Improve, and Control (DMAIC) steps are outlined in this book with examples given in cybersecurity. Noting that six sigma training requires statistics and technology knowledge for data analysis, this textbook covers the associated statistics, six sigma, lean, and technology-related concepts that a cybersecurity student or professional can use to an extent. Waste categories for Six Sigma's Cybersecurity application purposes is defined for the first time in this textbook with lean incorporated into the approach. This textbook can be appropriate for a lower or upper-level undergraduate course that depends on the statistics and technology requirements of the course. This book can also be used for six sigma certificate training of six sigma professionals and students.

Contents

1	**Introduction**	1
	Bibliography	13
2	**Kaizen Event & Six Sigma in Cybersecurity**	15
	2.1 Kaizen Event	16
	2.2 Six Sigma Projects	21
	2.3 Cyber Waste Categories	24
	2.3.1 Speed	25
	2.3.2 Access	25
	2.3.3 Resource Utilization	25
	2.3.4 Talent	26
	2.3.5 Utility (Effectiveness)	27
	2.3.6 Cyber-Transport and Motion	27
	2.3.7 Cyber-Production	28
	2.3.8 Superfluity Cyber-Processing	28
	2.3.9 Design Flaws	28
	2.3.10 Waiting	28
	2.3.11 Architecture	29
	Biobliography	31
3	**Statistics for Six Sigma**	33
	3.1 Probability	34
	3.1.1 Probability of a Continuous Random Variable	35
	3.1.2 Probability for Discrete Random Variable	36
	3.2 Basic Measurement Tools for Six Sigma by Using Basic Statistics	37
	3.3 Discrete Distributions	39
	3.3.1 Data Driven Discrete Distribution	40
	3.3.2 Discrete Normal Distribution	43
	3.3.3 Data with Binomial Distribution Nature	45

		3.3.4	Poisson Distribution	46
		3.3.5	Geometric Distribution	47
		3.3.6	Hypergeometric Distribution	48
	3.4	Continuous Distributions		49
		3.4.1	Normal Distribution	50
		3.4.2	Uniform Distribution	53
		3.4.3	Exponential Distribution	55
	Bibliography			59
4	**Let's Define**			**61**
	4.1	IPO Diagram		67
	4.2	SIPOC Diagram		70
	4.3	Problem Space		72
	4.4	Lean Six Sigma		73
		4.4.1	Customers	73
		4.4.2	Inputs	75
		4.4.3	Process	75
		4.4.4	Outputs	76
		4.4.5	Suppliers	77
	4.5	Cybersecurity Process		77
	4.6	Document Problems		81
	4.7	Flow Diagram (Chart)		83
	4.8	Business Process Mapping		85
	4.9	Functional Flow Diagram		86
		4.9.1	Project Charter	87
		4.9.2	Spaghetti Diagram	88
		4.9.3	Hierarchy Tree of Critical to Quality Conditions	91
		4.9.4	Initial Phase of Value Stream Mapping—Defining Operations	92
		4.9.5	Kano Model	95
	Biobliography			97
5	**Measuring Variables for the Defined Problem**			**99**
	5.1	Measure Basics and Statistics		101
		5.1.1	Project Measurement Planning	102
		5.1.2	Key Performance Measurements (KPM)	110
		5.1.3	Critical to Measurement (CTM)	113
		5.1.4	Measurement System Analysis Needs (MSA)	118
		5.1.5	Measurement Matrix	124
	5.2	Testing Hypothesis Statements		129
		5.2.1	One-Sample t-test	131
		5.2.2	Two-Sample t-test	131

	5.2.3	Paired t-test	131
	5.2.4	Chi-Square (χ^2) Statistical Method	133
	5.2.5	Development of Confidence Interval	135
	5.2.6	Confidence Interval Formula	137
	5.2.7	Choice of Sample Size	138
	5.2.8	Null Hypothesis	138
	5.2.9	Displaying Measure Phase Outcomes	144
	5.2.10	Takt Time (TT)	148
	5.2.11	Overall System Effectiveness	150
	5.2.12	Measuring Systematic Performance	152
5.3	General Software and Online Resources for Six Sigma Analysis		155
	5.3.1	C	156
	5.3.2	C++	156
	5.3.3	Java	157
	5.3.4	IBM SPSS	157
	5.3.5	Matlab	158
	5.3.6	Minitab	158
	5.3.7	Microsoft Excel	158
	5.3.8	Microsoft Visio	161
	5.3.9	Palisade Products—@Risk, Evolver, NeuralTools, PrecisionTree, StatTools, TopRank	161
	5.3.10	Python	161
	5.3.11	R and R-Studio	162
	5.3.12	SAS and SAS-Studio	163
Biobliography			164

6 Analyze the Measurements and System 167
   6.1 Five W and How Questioning 168
   6.2 Batch Means Method 170
   6.3 Constraint Optimization Analysis 172
   6.4 Correlation Analysis 173
        6.4.1 Positive Correlation 174
        6.4.2 Negative Correlation 175
        6.4.3 Neutral (no) Correlation 176
        6.4.4 Strength of Correlation 178
   6.5 Design of Experiment (DoE) Analysis 179
   6.6 End Goal Analysis 180
   6.7 Equal Variance Testing 182
   6.8 Failure Mode and Effect Analysis (FMEA) 183
   6.9 Interaction Table 184

6.10	Interface Analysis		187
6.11	Maximum System Capability Analysis		187
6.12	Overall Methodology Effectiveness (OME)		188
6.13	Pareto Chart Analysis		189
6.14	Process Efficiency Analysis		191
6.15	Right Lean Techniques' Selection		193
	6.15.1	Heijunka	194
	6.15.2	Jidoka	194
	6.15.3	Just in Time (JiT)	195
	6.15.4	Poka-Yoke	195
6.16	Resource Analysis		196
6.17	Root-Cause and Cause-Effect Analysis		196
6.18	SWOT (Strength-Weakness-Opportunities-Thread) Analysis		199
6.19	Time Analysis		201
6.20	Value Analysis		203
6.21	Value Stream Mapping (Revisited)		203
6.22	Regression Analysis		204
6.23	Analyze—Software		205
	6.23.1	American Society for Quality (ASQ)	207
	6.23.2	C	207
	6.23.3	C++	208
	6.23.4	Java	209
	6.23.5	SPSS by IBM	210
	6.23.6	MATLAB	211
	6.23.7	Microsoft Products	213
	6.23.8	Microsoft Excel	213
	6.23.9	Visual Basic for Applications (VBA)	214
	6.23.10	Add-Ins	217
	6.23.11	Minitab	217
	6.23.12	Palisade Products	217
	6.23.13	Perl	220
	6.23.14	Python	220
	6.23.15	R and R-Studio	221
	6.23.16	SAS and SAS-Studio	223
	6.23.17	SigmaXL	224
	6.23.18	Other Software Packages	224
References			226

7	**Improvement of the Systems**		229
	7.1	Introduction	230
	7.2	Software Improvement	231
		7.2.1 Design	232
		7.2.2 Duration	232
		7.2.3 Insight (Conceptual Coverage)	232
		7.2.4 Purpose	233
	7.3	5S (Five S)	233
	7.4	Network	236
	7.5	Improving Design	237
		7.5.1 Cyber Solution Design	237
		7.5.2 Transportation Design	238
		7.5.3 Document Design	238
		7.5.4 Duration	239
		7.5.5 Infrastructure	239
		7.5.6 Insight (Conceptual Coverage)	239
	7.6	Employee Training/Improvement	240
	7.7	Poka Yoke (Mistake Proofing)	241
	7.8	Simulation	241
	7.9	Data Analytics, Machine Learning, and Artificial Intelligence	243
	7.10	Time-Driven Improvements	244
	7.11	Do We Have Effective Improvements?	246
	7.12	SMED (Single Minute Exchange of Die)	248
	7.13	Did the Improvements Work?	249
	7.14	Value Stream Mapping—Phase of Improvement Applications	250
	7.15	Prevention of Mistakes	250
		7.15.1 Improve Through Controlling at Each Phase	251
		7.15.2 Improving Through Implementation of Control at Some Phases	252
		7.15.3 Improvements Without Any Control	253
	Bibliography		255
8	**Control for Sustainability of Improvements**		257
	8.1	Control Mechanisms in Cybersecurity	258
		8.1.1 Essentials for Setting up the Control Systems	259
		8.1.2 Organizational Control Mechanisms	261
	8.2	Control Mechanisms in Six-Sigma	266
	8.3	Designing Charts for Controlling	267
		8.3.1 Basic Definitions	267
		8.3.2 Designing Control Charts for Continuous Variable Data	270

	8.3.3	Control Charts for Discrete Data	278
	8.3.4	np Chart	279
8.4	Reacting to Out-Of-Control		282
	8.4.1	Operation Halting Mistakes	282
	8.4.2	Warning Requiring Mistakes	282
	8.4.3	Documentation and Finalization	283
References			285
Index			287

About the Author

Emre Tokgoz Ph.D., wrote this book based on extensive research, projects, and teaching experiences gained during his training of university students in Six Sigma, which included both Black Belt and Green Belt certifications. He collaborated with industry partners in the New York and Connecticut areas of the United States and also drew on his experience as a professor at the State University of New York's Farmingdale State College branch. Dr. Tokgoz completed two Ph.D. degrees, one in mathematics and the other one in industrial engineering. He also holds four master's degrees in computer science, mathematics, engineering management, and biomedical engineering, in addition to a Bachelor of Arts (BA) in mathematics. His research, publication, and teaching areas and interests included cybersecurity, AI and its applications, pedagogy, optimization, biomedical engineering, robotics, game theory, network analysis, financial engineering, facility allocation, inventory systems, queueing theory analysis, supply chain, renewable energy sources, STEM education, engineering management, and Riemannian geometry. This book is similar to the author's Six Sigma books for engineering technicians and industrial engineers, which are also part of the Synthesis Lectures on Engineering, Science, and Technology book series.

Introduction 1

Continuous improvement is an essential part of any organization with the goal of maximizing efficiency and reducing possible waste. The roles of Six Sigma and Lean Methodology in continuous improvement are to provide frameworks to organizations during such efforts. The ultimate goal is to increase user/customer satisfaction by improving products and/or services while minimizing costs and redundant/unnecessary activities that turn into waste. Cybersecurity is a field that can take advantage of both Six Sigma and Lean Methodology techniques' applications. This organizational improvement approach can help through waste reduction, quality improvement, delivery time efficiency, and cost reduction. The goal of this book is to explain Six Sigma and Lean methodology with its applications in cybersecurity. The definitions to be used throughout this work include the following:

- **Stakeholder/Customer**. Anyone who has an interest in the task completion.
- **Waste**: Anything a stakeholder/customer is not willing to pay for.
- **Work**: Any effort made to complete a product or service.
- **Value**: Anything a stakeholder is willing to pay for.
- **Value-Added Activities: (VA)**: Any activity that a stakeholder is willing to pay for such activity that adds value to the product or service.
- **Non-Value-Added Activities (NVA)**: Any activity that a stakeholder is not willing to pay for that does not add value to the product or service.
- **Business (or Enterprise)-Value-Added Activities (BVA or EVA)**: Any activity that is necessary to complete to be able to accomplish the end goal of the product or service. EVA and BVA will be used interchangeably throughout this book.

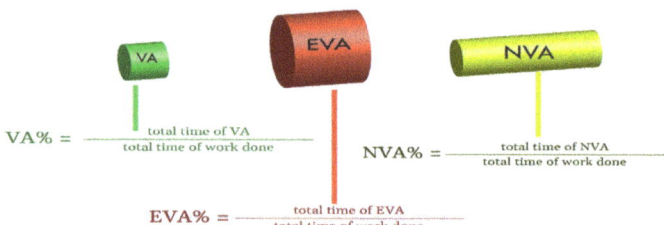

Fig. 1.1 Value added, non-value added, and enterprise value added activities' formulas

- **Input**. Any essential component or element needed to complete the product or service for any relevant work to be done.
- **Output**. The outcome of the completed product or service that the customer is expecting upon accomplishing the work by using the input.
- **Process**: Any activity that takes place between the input and the output of a product or service.
- **Lead Time**. The time between the beginning and ending of the process for completion of a production or service process.
- **Critical to Quality (CTQ)**. Anything that plays a role in the quality of the completed product or service. CTQ outlines the quality expectations of the stakeholder/customer to specify the goal of a Six Sigma project to fulfill customer's expectations.

We will use the term customer instead of stakeholder interchangeably throughout the book that can be considered the same from an organization's perspective. There are internal and external customers with internal customers being the customers that are members of the organization while the external customers are those who are external to the organization.

The definition of waste is very much stakeholder driven with the value analysis provided in Fig. 1.1 demonstrating how value added (VA), non-value-added activities (NVA), and enterprise value added activities (EVA) can be calculated in percentages by using the formulas.

The Greek letter σ (Sigma) is typically used for representing the mathematical standard deviation formula, and standard deviation helps us to measure by how much we deviate from the standard value (which is typically the average.) Six Sigma refers to being within 6 Sigma range from the average, and it is possible to mathematically calculate it. For instance, our production may be 150 items on average and deviation from this standard is determined to be 20. Hence, our Six Sigma range can be the numbers between 90 and 210. A Six Sigma project requires a walkthrough of the completed works and processes with evaluation of the inputs and outputs utilized for the achievement of the output. In such a walkthrough, it is eminent to face waste that results in the concept of a Waste

1 Introduction

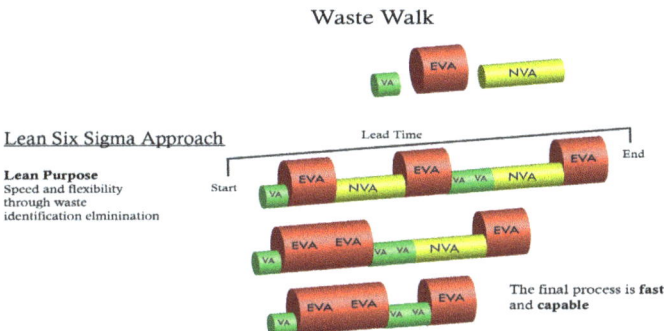

Fig. 1.2 Three steps of a project's walkthrough with the Waste Walk illustration displayed by the NVA

Walk which is a walkthrough of a work/process to identify waste that arises because of the NVA activities that took place in the completed work. A walkthrough of a completed task can be broken down into VA, EVA/BVA, and NVA activities as each step has a certain classification from stakeholders' perspectives. Such a walkthrough is demonstrated in Fig. 1.2 that outlines the VA, EVA, and NVA activities/work to be done (or already completed) during the completion of a project from the beginning to the end. This graph demonstrates all steps of the processes completed from the beginning to the end in a step-by-step fashion. On the top image, the lead time is shown from Start to End with the three EVA, three VA, and two NVA activities identified. The lengths of the icons can represent the duration of the associated activity when measured in minutes, hours, or days. For instance, the NVA activities in this image appear to be longer than the others therefore they are scaled proportionally from a time perspective. It is also possible to include other measures on such a graph that can include money, personnel, threads occurred, etc. The image below the top one in Fig. 1.1 shows that there is a reduction on the number of NVA activities from two to one with no changes in the EVA and VA activities; A Six Sigma project can result in such an improvement within the process. The final (bottom) Waste Walk shows elimination of the NVA activity existing in the process. Hence, at this final step, the process is fast and capable of producing based on the observations that align with the purpose of Six Sigma, Lean Methodology, organization, and customer's purpose: Improving the speed and flexibility of a product or service through waste elimination by eliminating NVA activities and maximizing efficiency. The breakdown of VA, NVA, and EVA needs to be implemented carefully, and it can depend on the experience and knowledge of the associated parties therefore a NVA considered by one person may not be considered as a NVA by another person. In fact, a process step considered as NVA in one activity may not be a NVA in another process. In determination of such works, experience matters and helping learners to gain background knowledge is one of the goals of this book.

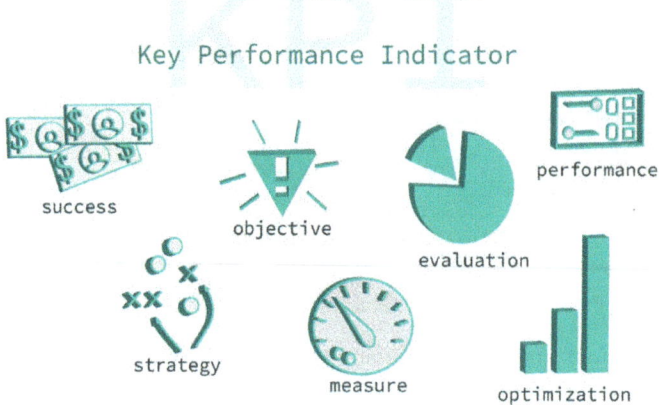

Fig. 1.3 A list of key performance indicator (KPI) that can be considered during a Six Sigma project

A Key Performance Indicator (KPI) is a term (that is not only associated with Six Sigma) that represents essential element of a system that play a key role in the expected performance of the system; such an element requires careful attention due to its impact on the system performance. The choice of KPIs during a Six Sigma project are likely to depend on the choices of the customer based on the expected satisfaction levels from the Six Sigma project. Such KPIs typically need to be measurable to derive comparative numerical results for tracking the progress and final outcomes of the project; A wide range of techniques can be used for comparing pre- and post-Six Sigma project measurements that include but not limited to statistics. Such results can be used later on for tracking possible changes within the system. Figure 1.3 contains several KPI components that can be expected during the implementation of a Six Sigma project:

- **Finances**. A project scope drives the KPIs from a financial perspective. There is always a trade-off between business aspect of the cyber-thread elimination and handling attack surface with the associated preventive measures with their costs, hence, in cybersecurity, and in other areas of Six Sigma project completion, not all projects directly aim to improve or reduce costs. For instance, in the case when a serious malware thread is eminent that can impact the entire organization and the external customers' personally identifiable information, a Six Sigma project can be designed to handle such a thread where the cost can be the least concern due to the CEO's interest in eliminating such thread as soon as possible. In some other cases, the costs may be hindered. For instance, the software inventory of a company used for eliminating cyber-threads can be organized in a way to minimize the software that does the same work to not only reduce the cost but also reduce the time it takes to run the two software packages.

- **Objectives**. Objectives of a project drive the KPI choices. For instance, if the goal of a Six Sigma project is to increase the quality of a system's design while keeping the total cost as low as possible for network security purposes then the cost reduction is the main goal of the project while keeping the existing quality of the operations stable as targeted by the customer. It is also possible to have multiple KPIs for a project. In such regard, maximizing the number of success thread responses, increasing the production quantity within a certain area of a facility, and minimizing the network intrusions may be the objectives of a customer with the associated three KPIs.
- **Measurements**. The choices of measurements for identification of success depends on the chosen KPIs with the consistency of the measurement units throughout the entire project. For instance, according to NIST's SP 800-53, passwords for service accounts are recommended to be at least 15 characters in length. Passwords may exceed 15 characters so long as Agency business processes are not negatively impacted. Therefore, if an organization decides to follow NIST's Agency guidelines then 15 characters may be normal however the organization may also choose to require its users to choose 12 characters of passwords.
- **Performance**. The performance of a project depends on the performance of all factors that relate to the completion of the project. Such factors can be multifold depending on the project on hand that can include employees, machines, integration of machines, and other factors. For instance, an automated Intrusion Detection System (IDS) may be provided by a vendor that relies on the system's functionality and decisions made to control the network security and the improvement depends on the desired outcome attained by the customers. The objective of the performance evaluation may depend on the speed of the IDS in detecting threads, the cost of the IDS, and its accuracy of the IDS in detecting threads.
- **Evaluation**. Any project evaluation is driven by the success metrics determined by the customer and the output's fulfillment that respects both the customer and the industry standards (if there are any.) The success and scope of the project is driven by the metrics used for KPI measurements based on pre- and post-project evaluations. For instance, if a certain thread has been measured to be 10.6 instances per day prior to the implementation of a Six Sigma project and post-project evaluation indicated the same instances to be 6.2 per day then it shows a measurable success trajectory as a part of the project's evaluation. The success of the project outcome depends on the expectations of the customer; For instance, if the customer stated the expectation of the instance to be reduced to 2.1 per day, then the project has some success however it requires furthermore improvements to be able to reach customer-desired success.
- **Strategy**. The strategy that can be followed for a Six Sigma project can be multifold and impact the project with the strategy of the KPIs, project's aim, goals etc. considered for the project. Noting that project improvement techniques and outcomes can be multifold, the results of a Six Sigma project would be limited to the strategy chosen. Hence, depending on the costs and resources, evaluation of 3–4 improvement strategies

can be ideal with the output evaluation depending on possible variational aspects of the inputs selected (if applicable.) For instance, if our aim is to improve the speed of a network security process, the strategy and KPI may vary as follows:
- *First improvement idea.* Focus on the employee aspect of the network security elements with the goal of reducing employee-made mistakes and how to handle them.
- *Second improvement idea.* Improving the automation aspect of network security to speed up the intrusion detection process.
- *Third improvement idea.* Focusing on human–machine interaction aspect of the processes for eliminating any waste that may exist. Talent waste can be one aspect that can be improved if talented individuals are not employed based on their talent levels within the organization.

All these project ideas can be implemented however it is a matter of the costs associated with the project to the organization and the stakeholders' interest in undertaking such projects; Therefore, the strategy should be chosen mindfully based on the time and customer expectations that also include time and budget constraints that may limit the associated decisions.

Optimization. Optimization can be applied anywhere with the aim of maximizing or minimizing certain aspects of the project, it can even be implemented mathematically by using formulas and constraints however we don't cover such mathematically intense concepts in this book. The ultimate goal of optimization is to minimize and/or maximize one or more objectives under certain constraints that are driven by the system needs and the associated performances. The application of the optimization approach during the DMAIC approach can be applied anywhere depending on the optimized work desired. For instance, finding the right levels of KPIs to be measured with the objective of maximizing CTQ conditions for the customer may be the essential target of a project. Constraints in such a case can rely on either the quantities or qualitative statements outlined by the customer. Hence, the listed customer desired expectations from the system with each specification of the system is essential.

Six Sigma is a management methodology designed to help organizations improve. Targets for improvement can be, but are not limited to business processes, products, and services. Applied continuously, Six Sigma is a constant examination of process with the primary objective of discovering and eliminating defects that detract from the operational mission. The Six Sigma methodology may be applied to any operation, and processes for managing cybersecurity are well within its scope. In this chapter, we introduce the Six Sigma framework for continuous quality improvement, and then model a generic cybersecurity program using the Six Sigma framework. We generally refer to the organization supported by a cybersecurity program as the "business." The business may be public, private, government, nonprofit, and/or have any combination of attributes as long as it includes an internal cybersecurity program that acts on its behalf. We provide examples of how to measure the effectiveness and efficiency of that program.

Note that Six Sigma is a methodology as opposed to a method. That is, it is a loosely defined process for pursuing improvement that is consistent with numerous variations of activities. The reference section of this book includes citations for some of the most popular methods of implementing Six Sigma. Business managers who adopt a Six Sigma approach are not by that fact held to any specific method. In fact, due to the wide variety of business goals and objectives, even best practices such as Six Sigma always need to be customized. This book describes a method of adopting Six Sigma that has proven useful in cybersecurity management, and we embrace the idea that our readers may learn from this approach and move it forward within their organizations in multiple directions.

There are a wide variety of cybersecurity management best practices and references section includes some of the most popular practices employed. Cybersecurity improvement processes and standards have been developed by a plethora of authoritative sources and risk management is essential in this approach. A cybersecurity framework (CSF) is a risk-based approach used for reducing cybersecurity risk composed of three parts:

- The Framework Core
- The Framework Profile
- The Framework Implementation Tiers.

Our research has determined that the influential example of the US National Institute of Standards and Technology's (NIST) Cybersecurity Risk Framework (CSF) provides the ideal level of abstraction for structuring problems faced by cybersecurity managers. In that document, cybersecurity processes are called "Functions" and each function is associated with a set of outcomes. From a systems engineering perspective, a function is a process that takes inputs and transforms the inputs into outputs. Functions have inputs and outputs as well as activate and exit criteria. NIST-CSF guidance emphasizes that function outcomes should be guided by an internal risk assessment that results in a "Target" Profile, defined as a set of outcomes needed to achieve the desired cybersecurity risk management goals. So, engineers who wish to implement the NIST Framework are confronted with a problem that looks like Fig. 1.4:

This book will provide guidance for sources of information that can be used to determine (i) whether cybersecurity functions produce desired outcomes, and (ii) whether an organization's "Target Profile" is fit for purpose. From a systems' engineering perspective, the first assessment is one of verification, and the second of validation. While a level of detail required to describe how Six Sigma benefits Cybersecurity is not articulated by NIST in its description of CSF Functions, we will rely on industry standard controls as described in more detailed authoritative standards document, such as NIST Special Publication 800-53, Security and Privacy Controls for Federal Information Systems and Organizations. We will also rely on standard descriptions of cybersecurity

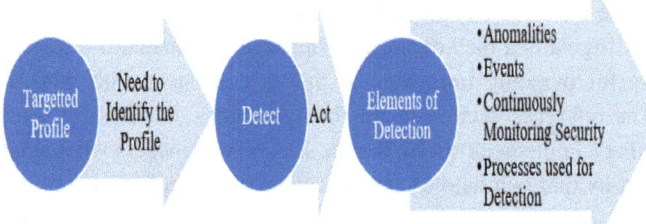

Fig. 1.4 A NIST-CSF function: "detect"

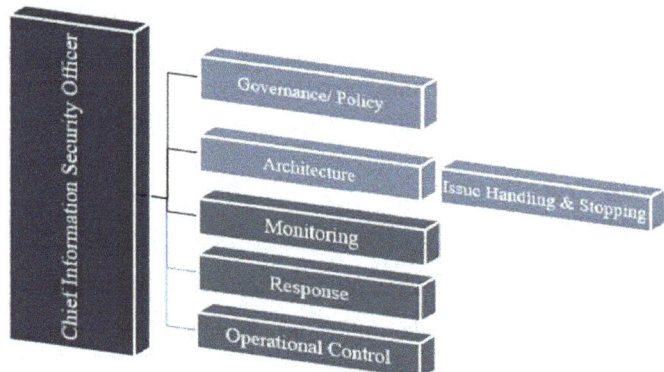

Fig. 1.5 Exemplar cybersecurity organization chart

program functions that appear in organization charts and technology control industry standard publication such as those issued by the Information Systems Audit and Control Association (ISACA). In very high-level narratives, we will fall back on generic names for controls provided by cybersecurity industry analysts such as those described in the TAG Cyber Annual Report.

Throughout this book, we will include examples of how the Six Sigma methodology may be productively applied to cybersecurity program management. Therefore, we start by defining what we mean by a cybersecurity program. Within any given organization, there is typically an individual assigned to manage any specialized function. In accounting, it is the Comptroller. In cybersecurity, the highest-ranking person in an organization whose sole job is cybersecurity is commonly called a "Chief Information Security Officer." That person will often delegate responsibility for establishing processes to achieve CSF functional outcomes via an organizational structure like the one in Fig. 1.5.

The responsibilities in figure example are typically delegated as follows:

Governance: Establish organization wide policies for maintaining data on the systems and information inventory, including requirements catalogue information and system operated by suppliers and business partners. This organization would also establish safeguards based on information classification.

Architecture: Design, develop, and deploy systems to enable the organization to comply with policy established by the governance function, to include controls for all technology and data inventory in the scope of established policy.

Remediation: An architecture function often includes security engineers who are responsible for acquiring technology or reconfiguring existing technology to contend with situations wherein policy non-compliance has been identified, or where known threats have not yet been considered by addressed by policy requirements. Though the engineers typically report to architecture, the organization chart places this function to the right of a Monitoring group to highlight the fact that it is typically on the receiving end of input from the monitoring function, wherein issues to remediate are often discovered.

Monitoring: A key control for any organization is a method of monitoring the environment to detect successful attacks, and/or highlight vulnerabilities that may enable future attacks. The org chart is labeled Monitoring / Response to differentiate it from administrative control responsibilities, but in practice this monitoring group is often called the Security Operations Center (SOC).

Control Operation: In small organizations, systems administrators will perform all security administration such as adding or removing users from system, configuring firewalls, etc. As organizations grow larger, these duties are typically segregated into one or more security control operations teams to minimize the number of people in an organization that have the ability to impact the integrity of security controls.

A team so composed generally includes the talent and leadership required to operate the NIST CSF Functions that include the following:

Identify: Develop an organizational understanding to manage cybersecurity risk to systems, people, assets, data, and capabilities.
Outcomes expected of Identify include Asset Management; Business Environment; Governance; Risk Assessment; and Risk Management Strategy.

Protect: Develop and implement appropriate safeguards to ensure delivery of critical services.
Outcomes expected of Protect include Identity Management and Access Control; Awareness and Training; Data Security; Information Protection Processes and Procedures; Maintenance; and Protective Technology.

Detect: Develop and implement appropriate activities to identify the occurrence of a cybersecurity event.
Outcomes expected of Detect include Anomalies and Events; Security Continuous Monitoring; and Detection Processes.

Respond: Develop and implement appropriate activities to act regarding a detected cybersecurity incident.
Outcomes expected of Respond include Response Planning; Communications; Analysis; Mitigation; and Improvements.

Recover: Develop and implement appropriate activities to maintain plans for resilience and to restore any capabilities or services that were impaired due to a cybersecurity incident.
Outcomes expected of Recover include Recovery Planning; Improvements; and Communications.

The responsibilities in the organization chart generally align to CSF functions as depicted in Figure X. There is some overlap because the Control Operation team is *responsible* for a variety of controls that cross all CSF functions, but the Figure shows which security team is *accountable* for the outcomes of CSF functions. Generally, the team appearing in Fig. 1.6 *beneath* the listed function is charged with enabling and maintaining quality control for the associated CSF function outcome. *Identify* involves Governance activities, *Protection* tools, standards, and techniques are provided by Architects, *Detection* is performed by the Security Operations Center (SOC), *Remediation* by Engineers *Responding* to events, and, as *Recovery* in NIST refers to the steady state where *Control* over operations is effectively maintained by Control Operations.

Cybersecurity programs generally define quality as the absence of negatively impacting events. But you cannot measure an event that does not exist, so to measure the quality of cybersecurity, one must measure attributes of the systems and threat environment and combine those measures in algorithms whose output reduces uncertainty about the likelihood of successful cybersecurity attack. Security monitoring and response typically follows a staged approach that begins with data collection. System and application user activity logs, automated alerts on configuration changes, and threat intelligence feeds are just a few of the measures that provide input to the cybersecurity detection function. That data is scanned for indicators of compromise (IoC). Suspected IoCs are forwarded to security event analysts as alerts to be manually analyzed. This initial scanning prompts the initial stage of a security monitoring and response sequence that includes:

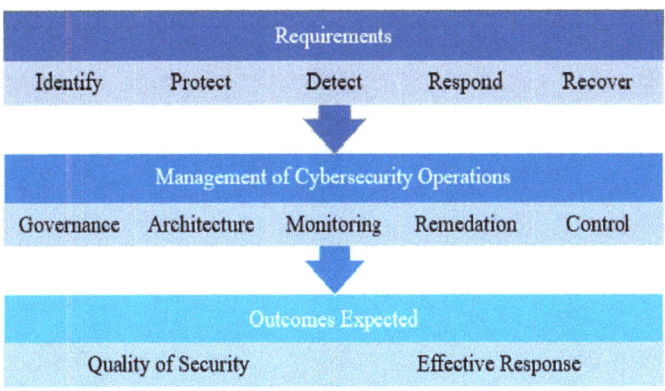

Fig. 1.6 Cybersecurity processes map to organizational functions

Identify: Review automatically generated alerts and the activities to determine if an attack is in place.

Analyze: Request additional data from additional sources and compile details on the series of systems' commands used for compromising the system. This results in a documented "threat vector."

Mitigate: If the analysis results in confirmation of a compromise of system security, a quality cybersecurity program will act to mitigate potential damaging impact.

Contain: Prevent the attack from expanding into currently unharmed systems by restricting the activities of the attacker via access controls.

Eradicate: Eradicate malicious activity and/or code from the impacted systems.

Information collected in each stage of incident response is then gathered for a post-mortem and root cause analysis, after which Response, Recovery and other CSF functions are often modified to incorporate lessons learned and improve overall program quality. Figure 1.7 shows the Cybersecurity Incident stages as subset of the Cybersecurity Program.

We have delved into the inner workings of the cybersecurity incident management subprocess to highlight the similarities in its pursuit of quality to the more general process quality improvement advice provided by Six Sigma.

Chapter 1 Exercises

Exercise 1.1 Give a specific example for Identify, Protect, Detect, Respond, and Recover from a cybersecurity setting/incident that may exist in real life. You can either provide a case that happened in real life (with appropriate citation) and explain it or research an application area that can be applied for such a case.

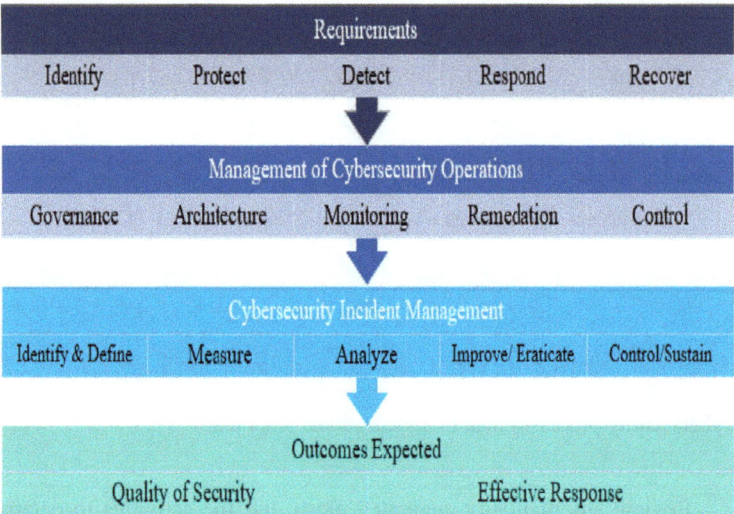

Fig. 1.7 Cybersecurity program map Six Sigma DMAIC

Exercise 1.2 Give a specific example for Identify, Analyze, Mitigate, Contain, and Eradicate from a cybersecurity setting/incident that may exist in real life. You can either provide a case that happened in real life (with appropriate citation) and explain it or research an application area that can be applied for such a case.

Exercise 1.3 If we need to identify KPIs for understanding possible threads as a part of phishing emails.

Exercise 1.4 In this section we covered NIST's recommendations for management of cybersecurity operations. Are there other organizations with similar sets of recommendations? List two possible organizations or management of cybersecurity methodologies other than NIST that can be used for management of cybersecurity operations. Explain how effective they are. Do they have requirement steps like NIST do?

Exercise 1.5 What are the responsibilities of a chief information security officer? How does such a person can benefit from Six Sigma applications?

Exercise 1.6 Explain possible VA, BVA, and NVA for IT personnel to handle phishing email decisions on whether the emails should be considered as junk or inbox emails. Explain how IT can decide such activities for email classification and judgement.

Bibliography

1. U.S. Department of Commerce, National Institute of Standards & Technology. https://www.nist.gov/itl/smallbusinesscyber/cybersecurity-basics/glossary
2. Tokgöz, E. (2024). *Quality and lean Six Sigma applications for industrial engineers.* Springer Nature. https://link.springer.com/book/9783031557392
3. Tokgöz, E. (2024). *Quality and lean Six Sigma for engineering technicians, synthesis lectures on engineering, science, and technology.* Springer, 978-3-031-44033-5. https://link.springer.com/book/9783031440328
4. NIST Computer Security Resource Center, Security and Privacy Controls for Information Systems and Organizations. Accessed January 22, 2025. https://csrc.nist.gov/pubs/sp/800/53/r5/upd1/final
5. Tokgöz, E. (2025, February). Artificial bee colony optimization techniques' utilization for intrusion detection systems' analysis. In *4th IEEE international conference on AI in cybersecurity (ICAIC) proceedings.* https://ieeexplore.ieee.org/stamp/stamp.jsp?tp=&arnumber=10848880

Kaizen Event & Six Sigma in Cybersecurity

Six Sigma is a method that provides tools for improving processes, products, and services with the goal of highest-level customer satisfaction. It was developed in the manufacturing industry, but it is widely applicable to any business processes, products, or services. The term business herein refers to any type of managed operation, whether for profit or not, public or private, partnership, corporation, or individual. Six Sigma is a method for waste examination of the business that detracts from its mission to serve its customers in the best possible ways. It prompts the identification of problems to be solved and applies a disciplined approach to problem solution. One way to apply six-sigma for any problem solution activity is known as "DMAIC" approach. This is an acronym for:

- Define
- Measure
- Analyze
- Improve
- Control

In this book, we use the domain of Cybersecurity to demonstrate how Six Sigma works. The demonstration also serves to provide a set of exemplary cybersecurity cases that may be adopted by an enterprise embarking on cybersecurity process improvement.

2.1 Kaizen Event

Even the most well defined and executed process must change as the technological landscape changes around it. Six Sigma DMAIC is a tool for continuous improvement, and it is a team-oriented approach that typically takes three to six months to complete. However, in the case of cybersecurity, the appropriate time to change process may be just moments after a cyberattack. In recognition that this type of scenario may occur in even the most well-run Six Sigma program, Six Sigma has defined an accelerated application of DMAIC, which is labeled a "Kaizen Event." A kaizen event engages all stakeholders within an organization to focus on a problem and its resolution.

In technology management, there is a mature thought process around DMAIC-like processes, and technology operations has itself been studied from a process improvement perspective. Perhaps the most widely recognized such effort stems from a publication by the UK Government's Central Computer and Telecommunications Agency (CCTA) in the 1980s: the Information Technology Infrastructure Library is a set of detailed practices for IT service management (ITSM) that focuses on aligning IT services with the needs of business. Now known only by the acronym ITIL, the library describes processes, procedures, tasks, and checklists which are not organization-specific nor technology-specific but can be applied by any organization to establish a baseline from which it can plan, implement, and measure technology services.

In ITIL, any interruption of the service, including a user request that may simply be a user error, may indicate the presence of an incident, problem, or known error. Incidents are not necessarily defects but can clearly be indicators of problems. Unplanned service outages are commonly labeled problems, as well as incorrect implementation of technology configuration standards. Figure 2.1 illustrates a standard ITIL Operations' Workflow. Six Sigma studies may easily adopt these definitions as a source of defect measurement and corresponding problem identification.

In the context of a technology operation, a cybersecurity event may surface as an unplanned outage, a user complaint, or an external report of internal data that has been posted to a data leak website. Security incident identification can be enabled by a technology incident identification process. An example of one such event is a ransomware attack. The first observation of such an attack is typically a service outage. For example, a business application may be unable to retrieve data from a server, or a user may be unable to access office files from a shared location. The incident will be classified as a problem because the cause is unknown. An administrator will be assigned to investigate. The investigation will reveal a ransom note from the attacker. The note may be posted at the root of a file system or displayed on an operating system console. It will inform the administrator that the data on the server has been encrypted and will also typically demand payment in bitcoin in return for the keys to decrypt the data and instruction on how to use them to restore the systems to pre-attack operational status.

2.1 Kaizen Event

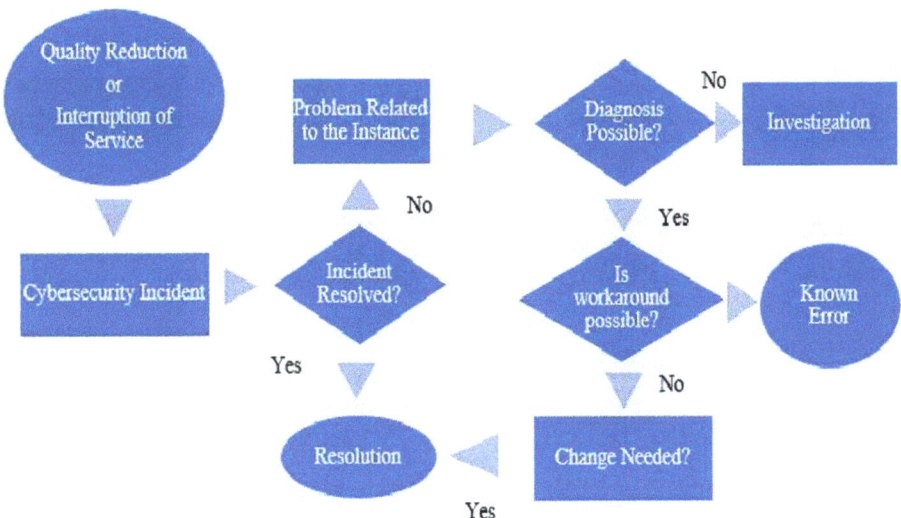

Fig. 2.1 IT operations workflow

The incident will be immediately classified as malware (*Malware* is a phrase attained from the words "malicious" and "software") attack and escalated to business management in a manner such as that displayed in Fig. 2.2. A SWAT team will be convened and the associated steps taken are outlined in Fig. 2.3. Like many cybersecurity terms, the acronym SWAT has its origins in the military. It stands for: "Special Weapons and Tactics," whether these be used in attack or defense. SWAT teams are formed by commanders who assign experts to represent them. The experts diagnose a situation and decide how to move forward. In direct response to cybersecurity incident identification, a SWAT team will be convened, and a decision will be made on how to protect assets and recover operations.

For this example, let us assume the SWAT team pays the ransom and restores the data. In conjunction with this activity, they decide to engage an external computer forensics response firm to quickly ascertain and document authorized data flow and assess whether the malware was introduced into the environment via an authorized or unauthorized process. Either way, they will implement network and operating system access restrictions on any data flow that could result in remote access capability to install additional malware. In the aftermath of this immediate containment response to such a ransomware attack, as well as of any other cyberattack, there will be a post-mortem to ensure complete eradication of any trace of malicious activity where the cybersecurity team embarking on cybersecurity process improvement adopted a six-sigma management approach, it would also immediately start planning a kaizen event.

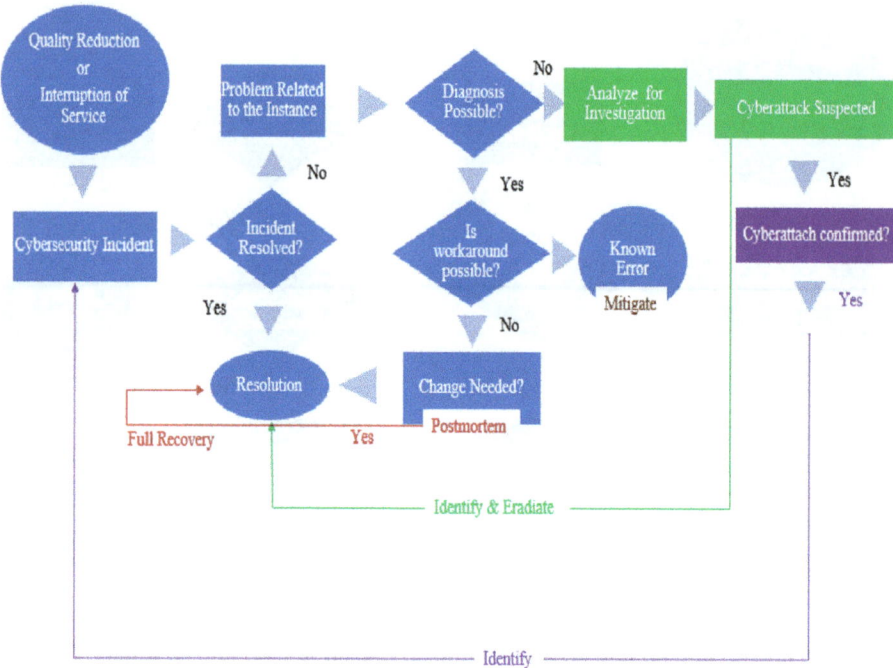

Fig. 2.2 Cybersecurity incident detection and response

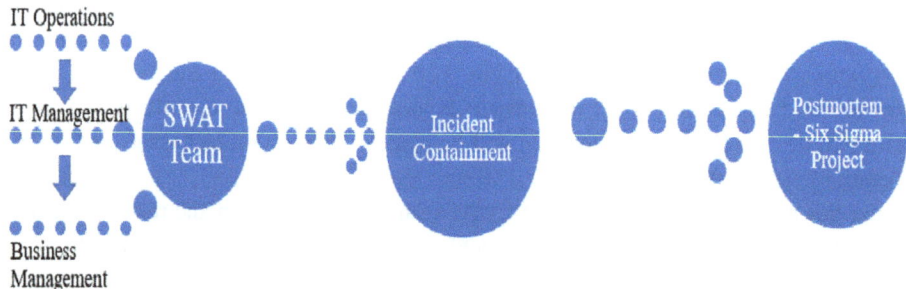

Fig. 2.3 Escalation of incident handling in the organization

This kaizen event would be a combination post-mortem and process improvement team in which everyone present has some knowledge and/or expertise to contribute that will enable the group to fully diagnose the root cause of the attack, and design corresponding process improvements to ensure that the root cause is remediated so that the next attack of the same event type will be unsuccessful. The kaizen event would follow all the steps of Six Sigma using an accelerated approach with the goal of eliminating the vulnerability that allowed the attack to occur. The steps would be:

- Define—Document problem and underlying proximate and root causes, as well as the data available to support the problem definition
- Measure—Develop algorithms, charts, and other methods of data presentation
- Analyze—Adopt and employ a structure approach to analyze the data
- Improve—Implement controls suggested by analysis results
- Control—Enhance process monitoring to cover control improvements.

Cybersecurity practitioners who have participated in cyberattack post-mortems are familiar with a large part of the kaizen drill. It is typically an in-person exercise in front of a large white board whereupon the attack vector is sketched, and the underlying evidence for each step in the vector is described in a manner that makes clear whether controls designed to prevent each step were planned, and if planned, implemented, and if implemented, correctly implemented. There is typically an evidence repository readily available via computer screen projection so that actual evidence may be examined in detail by the groups, as questions are raised and answered, and various theories of proximate versus root cause are proposed, tested, and adopted. As it is well understood by the group that each step was taken by the attacker to complete the attack vector, it becomes clear whether planned controls verified to be correctly implemented have failed validation.

Even cybersecurity practitioners who have not participated in such an exercise will recognize its theoretical basis in a threat analysis. There are online repositories of common attack vectors used by such as the MITRE ATT&CK® knowledge base of threat actor tactics and techniques based on real-world observations. These provide a starting point for assembling a custom "attack tree." The threat, or *goal* of an attacker, is at the root of the tree. Alternative sub-goal combinations that would achieve the goal are progressively decomposed until a combination of a few actions is sufficient to execute the threat. In a theoretical exercise, the SWAT team will devise alternative protection mechanisms that will reduce the probability that the threat actor's goal will be achieved. The difference between a theoretical exercise focused on improvement and an actual attack postmortem is that a postmortem uses available evidence to figure out which branch was traversed rather than just demonstrate the plausibility of their enactment.

In any cybersecurity postmortem exercise, and the hastily assembled gathering of experts will be focused on ensuring any remaining attack vectors are highly improbable. The team is briefed on the facts that an attack occurred and looks for evidence that the most plausible of the alternative methods by which it may have been enacted. In the kaizen version, the cybersecurity team will chart out the actual attack vector as part of the problem definition or *define* stage. They will then gather data on their own response to the attack. For example, they may chart their time to respond to the event once it has been identified. During an incident, SWAT team will record the time at which the response activities outlined in Fig. 2.3 were taken.

Encryption of a victim's (important) files in demand of a payment (ransom) to restore access is called Ransomware malware. Ransomware victims receive a decryption key only if the ransom payment is made. The malicious actor publishes the data on the dark web or blocks access to the encrypted file in perpetuity in case when a payment is not made. A Ransomware attack can be designed as a tree to analyze such an attack. There are two main levels/pathways that can be executed to be able to design a Ransomware attack:

- Operating system
- Storage device

The use of these two levels can result in some of the following outcomes depending on the conditions:

1. Use an insider
    1.1 Corrupt an insider
    1.2 Insider executes malware
    1.3 Instruct users on use of available tools
    1.4 Bribe insider
    1.5 Social engineer the insider
2. Use an authorized network path
    2.1 Gain remote access to the network path
    2.2 Gain remote access to exploit the edge software vulnerabilities
3. Use software vulnerabilities
    3.1 Exploit operating system vulnerabilities
    3.2 Encrypt as an authorized user
    3.3 Exploit edge system vulnerabilities
4. Encrypt as an authorized user (an employee or management console)
    4.1 Use phishing credentials of an authorized user
    4.2 Access management console after encrypting as an authorized user

Figure 2.4 is an example of how such measures are viewed. The top half of the Figure shows the measure and the lower half, a corresponding metric wherein several identified cybersecurity incident response measures are combined. Both parts are expected to be used for the analysis during the postmortem. Such data would be combined with incident details to shed light on potential improvements. For example, to see if a process or procedures were in place to support investigation and resolution activities, and/or if similar bottlenecks may be preventing quick resolution of incidents. Six Sigma applications on such data require analysis of the data to further Identify, Analyze, Mitigate, Contain, and Eradicate related activities in relation to the DMAIC approach for further improving the associated steps. Noting that it is a continuous improvement process, a graph similar to Fig. 2.4 can be attained several times during the improvement process to attain measurable outcomes upon the changes made on the cybersecurity systems.

2.2 Six Sigma Projects

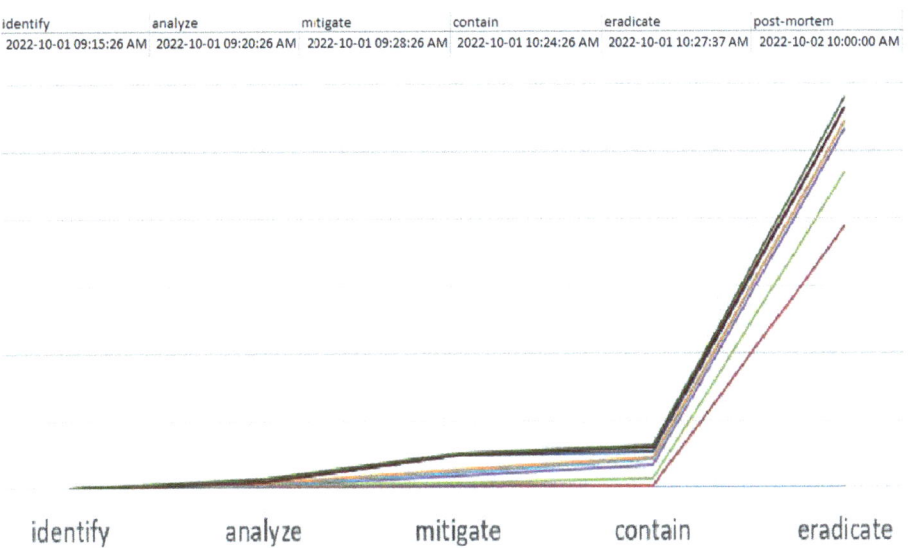

Fig. 2.4 Example of a Cybersecurity time to respond measurement

2.2 Six Sigma Projects

The full set of activities in kaizen Six Sigma kaizen event are the same as that on any Six Sigma project, just on a smaller scale with tight timelines. These are:

- Define—Document the attack vector and the vulnerabilities exploited to enact the threat.
- Measure—Collect data on system access controls and corresponding configuration variables intended to control.
- Analyze—Identify the gaps in control that allowed the attack vector to be traversed via unauthorized activity. Publish all relevant pre-established measures such as the time to respond to data 5. Gather all system activity logs and search them for evidence of the attack. Determine whether alerts on that activity would have allowed operations to stop the attack. List gaps in the logs.
- Improve—Devise new controls or enhance existing ones to prevent unauthorized traversal of the attack vector, to alert on such unauthorized attempts, and to respond to them.
- Control—Devise a method of monitoring the configuration of the system controls that would signal if they were not correctly configured or correctly functioning. For example, tripwires on the configuration variables and periodic automated vulnerability scans (Fig. 2.5).

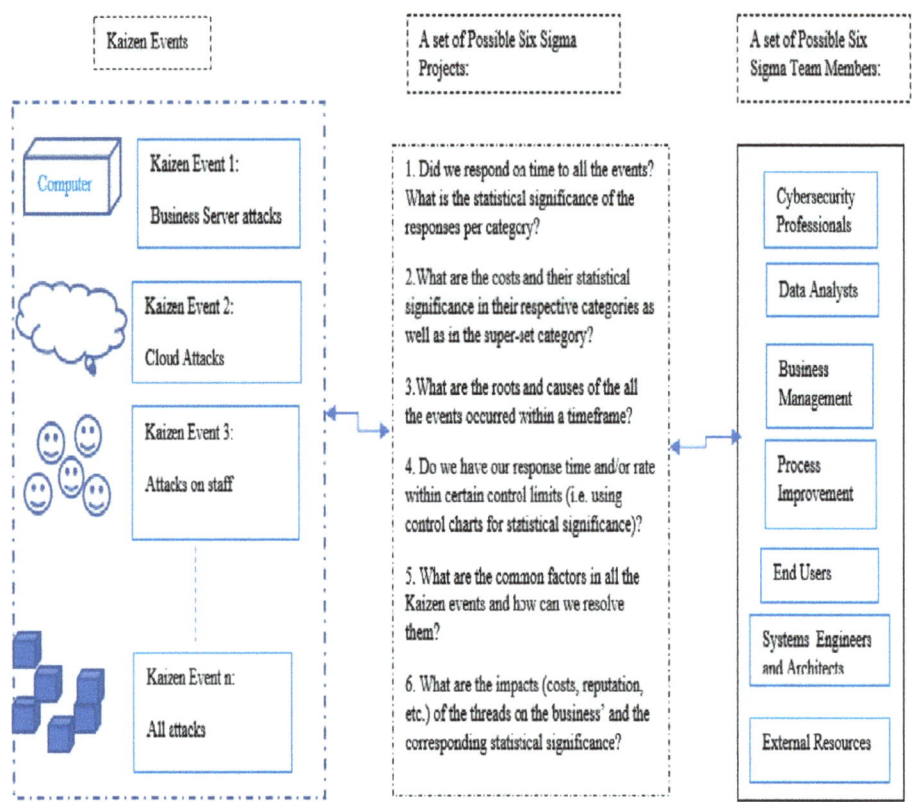

Fig. 2.5 Outline of a Set of Possible Six Sigma Projects

The main goal of the Six Sigma project is to incorporate statistics into business processes' improvement for business excellence. In the case of Cybersecurity incident management, such statistics are assembled one incident at a time. They also may be created with simulated incidents, known in the cybersecurity field as red/blue/and purple team exercises, and using hypothetical tabletop exercises, known in the field of operational risk management as scenario analysis. The aim is to relentlessly reduce waste and redundancies to increase productivity and efficiency. It is important to see the "big picture" of the operations in place and connect the outcomes to have a "wholistic" solution to the existing problems by gathering experts in the corresponding fields. The Kaizen events may be combined with improving responses to specific types of events as well as the response process.

- Kaizen event category 1: The set of cyber-attacks occurred to business application servers in the past three years and the corrective actions taken to resolve these issues with the corresponding costs.

2.2 Six Sigma Projects

- Kaizen event category 2: The set of cyber-attacks on cloud-hosted assets in the past three years and the corrective actions taken to resolve these issues with the corresponding costs.
- Kaizen event category 3: The set of cyber-attacks on employees' personal computing devices that can impact business operations in the past three years and the corrective actions taken to resolve these issues with the corresponding costs.
- Kaizen event category n: The superset of cyber-attacks in all other Kaizen event categories throughout the last fiscal year and the corrective actions taken to resolve these issues with the corresponding costs.

The above-mentioned groups of the Kaizen events can be analyzed in the context of a larger Six Sigma Project strategy by using the DMAIC approach using statistical analysis and significance. Good questions to ask about these sets of events include, but are not limited to:

1. Did we respond on time to all the events? What is the statistical significance of the responses per category? This project would require determining the success of the response to the events and comparison with the corresponding expected response time. The corresponding costs with cause-effect analysis and control charts can help to determine where the excellence of the overall responses stands.
2. What are the costs and their statistical significance in their respective categories as well as in the final all-inclusive category? Do we record enough data for events in any category to make comparison between event categories meaningful? If we consider all (i.e., n number) Kaizen events we listed above then the data corresponding to each Kaizen event category can be processed as an individual batch and analysis of the overall responses can be analyzed by using the batch means and other batch-based methods along with the control charts for determining whether the responses are under the expected control or not.
3. What are the root causes of each of the events that occurred within each year in the three-year timeframe? This requires looking at all the Kaizen events and seeing the big picture of all the responses and looking into the effectiveness of the corrective actions taken to be more mindful about the future actions to be taken.
4. Have we targeted response time and/or rate to fall within certain control limits (i.e., using control charts for statistical significance)? The responses to cyber-attacks with the corresponding cost can be analyzed and might appear to be out of control from both money and time perspectives. An improvement can be changing the cyber-defense structure and working with an external company in addition to the internal problem handling.
5. Are there common containment fail factors in all the Kaizen events and how can we resolve them? This would be the analysis of all the issues that led to the Kaizen events and seeing the big picture for corrective actions.

6. What are the impacts (costs, reputation, etc.) of the threads on the business and the corresponding statistical significance? A Six Sigma project to determine the trade-off between multiple measures observed in the combination of all Kaizen events.

All the above-mentioned Six Sigma projects require a set of professionals to gather and input data and ideas to be able to analyze and improve the problem in place. The members of the Six Sigma team can be from the following groups:

- Cybersecurity professionals—Can be contracted or employee of the company; could be a combination of both.
- Data Analysts—The Measure and Analyze phases of DMAIC require extensive data analysis therefore a person with data analytics skill sets should be present if none of the cybersecurity members have robust data analytics experience.
- Business Management—The "champion" or the sponsor of the project needs to be involved in the project to provide oversight to ensure alignment with business goals and objectives, as well as to be the decision maker in cases where the team may be hesitant or deadlocked. Process Improvement Staff—These contributors can come from any job function, but must be empowered to propose, document, socialize, implement, and test changes to processes, standards, procedures, and measures. Such contributors are key resources required to ensure the success of the DMAIC approach. If missing from a Six Sigma project, then the project application may stall at a theoretical level.
- End Users of the systems in scope of the cybersecurity incident impact are also key resources, and their feedback on whether systems are positive or negatively impacted by the change is a key measure.
- Systems Engineers and Architects—These members can help the team to see the big picture of the systems in operation and thereby to improve all members' ability to analyze the systems.
- Contracted-External Company—Depending on the scope of the project and the experience of external staff with Six Sigma, it may also be helpful to engage an independent consultant or company to provide data analysis and advice. This member can be particularly helpful in the improvement phase to introduce quality improvement techniques with which the company may not be familiar.

2.3 Cyber Waste Categories

In this section we list the waste categories that can be seen in cyber systems and provide examples to classify these waste categories so they will be useful as data categories in Six Sigma improvement projects.

2.3 Cyber Waste Categories

2.3.1 Speed

Tuning the speed of a cyber-solution until it reaches the optimal level is critical for many reasons to decline waste. The speed for accessing information, responding to threats, accessing the database, and finding the right response method to the corresponding threat, the speed of the system to communicate are all important considerations. If the system acts too fast, then the resources might be under-utilized in the system and a wrong decision might be made. If the system is too slow then the response to threats and information access could slow down which would also be issues; therefore, finding the right (i.e., optimal) speed of cybersecurity-related actions is the key along with the resource utilization. Hence, a high-speed cybersecurity solution might be missing too many critical elements to handle the threads while a low-speed solution might result in much more critical and harmful results in handling threads. Simulation of the designed process can help with finding the right tune of the process for reducing waste.

2.3.2 Access

Access utilization or permissions given to individuals, software solutions, or machines in cyber security include email, databases, resource, software, hardware, and cloud access. Under and over access to any one of these external and internal digital and physical environments can cause waste due to the increase in the vulnerability of the security system. There is an argument that over-access in technology authorization is a waste of technology resources generally and introduces vulnerabilities. For example, people who work in the mailroom do not generally need access to Internet email, but economies of scale in technology often provide cookie-cutter services where Internet access is included by default in everyone's email. This is a source of waste in technology that has consequences for cybersecurity, and so measuring technology waste is a good security metric—that is, any access by anybody that is not needed for job function.

2.3.3 Resource Utilization

There are several categories of waste from a resource perspective. These include, but are not limited to:

- **Internal computations.** Includes physical and digital resources such as data storage, internal software developed, the software resources used for cyber threat detection and prevention internally, the internal physical and digital media, the internal accounts used by the organization for accessing information.

- **External Computations.** Many companies do not regularly monitor and remove staff access to cloud providers, so daily or hourly comparison of cloud user lists with internal IAM systems is a good metric, and any users of cloud system not in internal IAM systems (fired, left etc.) is a good metrics, also usage records by these users at times when they are no vacation may indicate account compromise.
- **Storage.** Space is always an issue because business owners are overly cautious about purging dormant data. A strong record retention policy with automated archiving and delete processes could measure as well as reduce waste in this category to an acceptable risk tolerance level. Storage location matters: The data that is accessed the most needs to be in the most easily accessible location and ease of access can be calculated by using the basic probability theory. The hardware used for storage is vulnerable to cyber threads. The hardware where the information is stored can be accessible both physically and online. The method of storage needs to be optimized in such a way that the number of connections and the number of security measures should neither be overwhelming for the users nor be open to physical and online threats. We can group this storage into two different categories:
    - **Cloud Storage.** Cloud based storage is an external method of data storage while the local storage is storage without any net connection. Optimal levels of storage is crucial for fast and safe access to data. The following are important aspects of storage waste elimination:
    - **Local storage.** Utilization of the local storage with the data that is required to be accessed the most. This is particularly important for fast and reliable access to the stored data. This approach also reduces the probability of vulnerability to threads if the local security system is strong.
- **Acceptable Usage.** Some overly permissive acceptable usage policies allow staff to use business devices on social media sites that are crawling with malware. They may also be authorized to conduct personal financial transactions on business devices and to store large personal data objects like pictures and videos. All these policies introduce potential business vulnerabilities because threat actors targeting employees end up getting access to business systems. Note that these policies have their roots in the legacy environment wherein email was a mysterious commodity so there is little risk to employee cultural comfort in changing them now.

2.3.4 Talent

The right use of the talent to implement the right cyber security applications is crucial to avoiding waste. It is all too common to assign skilled cybersecurity analysts to routine monitoring of automated systems in situations where a smart high school graduate could easily meet the challenge, preserving the skilled labor for responses where more professional judgement is required. It is also common to use post-mortems of security incidents as large gatherings as if they were training sessions, when it is far more efficient to have a small set of more highly skilled experts render decisions that can be translated into rote training programs for less mature staff without sacrificing knowledge transfer.

2.3.5 Utility (Effectiveness)

Utility is like a resource, but utility waste is not directly correlated to the purpose of utilizing a resource but also considers whether the resource itself could be used more effectively. Categories of utility waste include, but are not limited to:

- **Machine learning curve.** This is where Artificial Intelligence algorithms are employed to spot the proverbial needle in a haystack, the computational resources allocated to this function should be closely compared with the cost–benefit of a more direct if–then rule-based incident identification. The learning curve is a machine learning technique used for reporting learning progress and performance on studying a specific method. A low rate of learning ability on the learning curve can be a waste due to slow performance of the system or more efficient direct methods of incident identification.

2.3.6 Cyber-Transport and Motion

- **Cyber-transport** is grouped into two subcategories: (1) Transporting of information and (2) transport of physical entities. The following areas are possible transportation waste:
  - ITS Architecture and Standards Security—Focuses on the development of architecture and standards required to ensure security in the connected vehicle environment.
  - Standards Overlap. There is waste in redundant standards—engineers deciding individually per area of responsibility broken down by regional or department affiliation rather than technology.
- **Cyber-motion** is the category of unnecessary authenticate or useless control measures because they can be easily defeated if not fir for purpose—captchas used for identification when threats are not robots, but threat actors with plenty of time on their hands. The number of connections in a network can be a burden on security applications and therefore NVA connections can be eliminated to have an optimal cybersecurity platform.

2.3.7 Cyber-Production

Production can be over or under the expected optimal product specifications therefore it causes waste. The optimal product is the one that matches the cyber specifications of the customer that matches with the CTQ conditions. The customer defines the CTQ conditions, however the cyber security professional needs to be mindful about the additional quality matters that need to be reminded of to the customer. Finding the optimal product might require several iterations and re-iteration of multiple versions of the resulting product based on customer's experience.

Over production and process waste from generating "ScoreCards" or dashboards not used for operation, but only for demos of technology when guests tour the security operations center.

2.3.8 Superfluity Cyber-Processing

This is the type of waste that happens when the optimal level of processing is not implemented. In this case the two types of process waste can be over-processing and under-processing. Over processing can be a result of repetitive or redundant process steps. Under-processing can be a result of missing critical steps in the process that are essential for the success of the process. This type of processing can result in many different issues related to money, time, energy, and speed related issues. Manual compilation of metrics by having engineers extract and send extracts to a security team who manually uploads them instead of automating data imports on a scheduled basis are typically done for assessments and audits, sometimes for monthly or quarterly reports.

2.3.9 Design Flaws

The design of a cybersecurity system may cause flaws within the overall system. Such flaws would be determined by the gaps that exist in the system and these gaps are identified by observing the overall system with possible issues to be detected within the system.

2.3.10 Waiting

Waiting can be a waste if the input, process, or output doesn't fulfill the purpose of the outcome. The "wait", depending on where it happens, can be a VA or BVA in which cases it is not assumed to be a waste. If it is determined to be an NVA by the customer or experts on the system, then it becomes a waste. For instance, a cyber thread detection

2.3 Cyber Waste Categories 29

mechanism waits for too long to respond to a thread based on the response rate in the system then it becomes a waste if it is known that there are better ways to handle it. The customer might be expecting a quicker response by the system therefore it would require continues improvement of the mechanism to respond to threads. The initial model may need to be improved for further progress in the response rate to threads.

2.3.11 Architecture

- Cybersecurity Standards—Focuses on the development of architecture and standards required to ensure security in the connected vehicle environment.
- STANDARDS OVERLAP There is waste in redundant standards—engineers deciding individually per area of responsibility broken down by regional or department affiliation rather than technologic Cyber-defects. A cyber defect can occur from a software perspective if the planning of the software development is not accomplished properly. In this case we can have a "reject" for the software developed if it doesn't fulfill all the requirements of the cyber needs that the customer is looking for. This reject can be reworked if
- Cyber-design. Two important aspects of a cybersecurity solution are sustainability and reliability from a software perspective. A design that is difficult to modify, justify (based on changing threads) and improve for additional threads could cause waste and result in not only problems faced for security but also would result in money and time waste. Six Sigma application can be particularly important for reducing cyber-design waste by seeing the "big picture" in the system in a wholistic perspective and act upon it through developed Kaizen events if needed. Sustainability during the control phase of DMAIC becomes an important part of the reduction of cyber-design waste.
- Software Acquisition: Using the right software to respond to the right threat and developing the right software for the right purpose can be listed in this category. Acquisition is a huge source of waste due to confusion on the part of buyers with respect to the interoperability of cybersecurity software features and functions.

Examples of areas where Six Sigma can be applied to cybersecurity include but not limited to the following:

- Vehicle Cyber Security—Focuses on preventing attacks from entry into our vehicle systems and components.
- Infrastructure Cyber Security—Focuses on protecting against threats and vulnerabilities to a nation's roadside equipment, devices, and systems.
- Dedicated Short-Range Communications (DSRC) Security—Focuses on ensuring trusted communications between vehicles and between infrastructure and vehicles.
- Security Credential Management System (SCMS) Operations

Example 2.1 Host-Based Intrusion Detection (HIDS) is a specialized layer of security software that is used for analyzing possible threat traffic and logging malicious behavior. It is simply an intrusion detection system that monitors the computer infrastructure on which it is installed for visibility of what's happening on your systems' critical security. Upon monitoring activities to detect suspicious behavior by using HIDS, the following can be identified:

- Intrusions
- Logged suspicious events
- Sent alerts
- Monitored system calls
- DLL activity
- External and internal intrusions

A Network-based Intrusion Detection System (NIDS) analyzes possible threat traffic and logging malicious behavior on a network. Information logged (i.e., NIDS) by NIDS sensor includes the following:

- Timestamp
- Connection or session ID
- Event or alert type
- Rating
- Network, transport, and application layer protocols
- Source and destination IP addresses
- Source and destination TCP or UDP ports, or ICMP types and codes
- Number of bytes transmitted over the connection
- Decoded payload data, such as application requests and responses
- State-related information

Depending on the scope and nature of the issues faced, there can be a variety of Kaizen events and six sigma projects that can be designed. An example of a six-sigma project can be to focus on malicious activity rate and number of bytes transmitted through connections with the actual number of issues occurred during these times and how the associated issues can be resolved by identifying the need for increasing the level of security during such activities.

Example 2.2 Examples that may be waste or cause waste include the following:

- Vulnerabilities
- Codebase weaknesses
- Design flaws

- System configuration weaknesses
- Architecture
- Authentication
- Privilege changes
- Database access models
- Mobile code
- Design of manual security
- Configuration gaps
- Software gaps

There are vulnerability scanners that can help to close the gap caused by code and design flaws.

Chapter 2 Exercises

Exercise 2.1 Would it be possible to form a cybersecurity six sigma event as a combination of Kaizen events? Please explain the reason for your response.

Exercise 2.2 What type of wastes may occur upon the breach to a university upon a phishing email opened by the president of the university.

Exercise 2.3 List and explain five wastes that may occur upon a weakness of Host-Based Intrusion Detection (HIDS) system. Recommend a mitigation strategy for elimination of such weaknesses.

Exercise 2.4 Find a real-life cyberattack case in which at least five wastes occur. List and explain how these wastes occur in this application.

Biobliography

1. Lisanti, Y., Luhukay, D., & Mariani, V. (2017). IT service and risk management implementation for online startup SME: case study: online startup SME in Jakarta. In *2017 International Conference on Information Management and Technology (ICIMTech)* (pp. 300–303). IEEE.
2. Dybdahl, S. O., & Staer, M. N. (2023). *Designing a framework for data populating alarms based on mitre techniques.* Master's thesis, University of Agder
3. Tokgöz, E. (2024) Quality & lean six sigma applications for industrial engineers. Springer Nature, Switzerland. https://link.springer.com/book/9783031557392
4. Tokgöz, E. (2024) Quality and lean six sigma for engineering technicians. In *Synthesis Lectures on Engineering, Science, and Technology.* Springer Cham 978-3-031-44033-5. https://link.springer.com/book/9783031440328

5. Tokgöz, E. (2025). Artificial bee colony optimization techniques' utilization for intrusion detection systems' analysis. In *4th IEEE International Conference on AI in Cybersecurity (ICAIC) proceedings*. https://ieeexplore.ieee.org/stamp/stamp.jsp?tp=&arnumber=10848880

Statistics for Six Sigma

The representation and analysis of data requires the use of statistics; therefore, statistics is conceptually the most important unifying concept of all measure, analyze, improve, and control phases. There are two well-known types of statistics, descriptive and inferential, that can be used as a part of measurements. The *descriptive statistics* represent the statistical results based on the period that the data is collected. We want to highlight the "the time period that the data is collected" because the period of the data matters; what is measured a month ago may not necessarily represent today's system. Inferential *statistics* is used by which observations are made based on the existing data to extract meaningful outcomes for future observations. The periodic observations and the seasonal considerations of the data also need to be kept in mind during the data collection and the type of statistic used during measurements.

In applications, a very basic level of statistics such as average (i.e., mean), standard deviation, mode, and median values are used often. These concepts are followed by statistical concepts such as confidence interval, distributions, and hypothesis testing. More advanced concepts such as Central Limit Theorem, Law of Large Numbers, hypothesis testing, advanced normality tests, capability analysis measurement, Gage R&R (for both Discrete and Continuous Data) are also covered.

We state the formulas corresponding to these and many other statistical measures used in Six Sigma in this book however, in many applications, it is unnecessary to know these mathematical formulas and you are expected to understand the meaning behind these formulas. For instance, Excel functions are easy to use for determining values for such formulas and we will explain these functions as we advance in the concepts. Other programming paradigms such as C++, Java, Phyton and MATLAB have their own functions for those who are interested in using these computational tools.

In this section the following will be covered:

- Random variable (both discrete and continuous random variable).
- Probability density function.
- Mean, mode, median, variance, standard deviation, range, and confidence interval.
- Exponential and uniform distributions with examples and figures.
- Continuous and discrete normal distributions with examples and figures.

There are certain events where the "chance" is involved in calculations. For example, when we flip a coin there is a chance of getting a head or a tail. The chance (probability) of getting a head or a tail after one time flipping coin is 0.5 since there are only two choices as possible outcome and one of them must hold.

Random Variable. The events for which we determine the probabilities are called random variables.

- **Continuous random variable**. An event/variable with real number outcomes.
- **Discrete random variable**. An event/variable with discrete outcomes.

Example 3.1 Time is a continuous random variable because it can be any real number while the number of threats detected is a discrete variable as it has discrete nature.

Statistical considerations such as probability, mean, and standard deviation have a major role in calculations and concepts taking place in measure, analyze, improve, and control phases. These calculations depend on the nature of continuity or discrete nature of the random variable that will be explained next.

3.1 Probability

Probability. Probability is a measure used for determining how likely an event can happen when all the possible outcomes are considered. The probability p(x) of an event x's happening is between 0 and 100%. i.e., $0 \leq p(x) \leq 1$.

- If p(x) = 0, then it is impossible for event x to happen.
- If p(x) = 1, then event x is certainly going to happen.
- The probability of a random variable can be
  - Discrete.
  - Continuous.

Distribution. When we consider a specific event and all its possible outcomes, the change in the outcome follows a specific pattern called the distribution of the quantity considered.

3.1 Probability

- **Continuous distribution.** A continuous distribution is the distribution of a continuous random variable that forms a continuous function that we will call **probability density function (pdf)**. The nature of a pdf's graph has a continuous structure just like the images shown below.

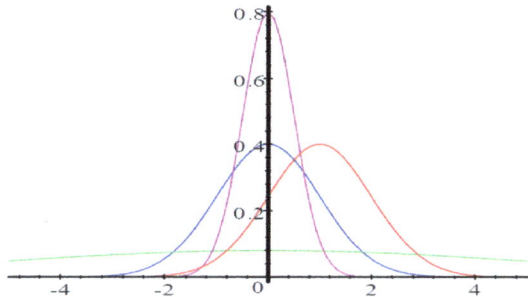

- **Discrete distribution.** A discrete distribution is the distribution of a discrete random variable that forms a discrete function we will call **probability mass function (pmf)**. The nature of a pmf's graph has a discrete structure just like the image below.

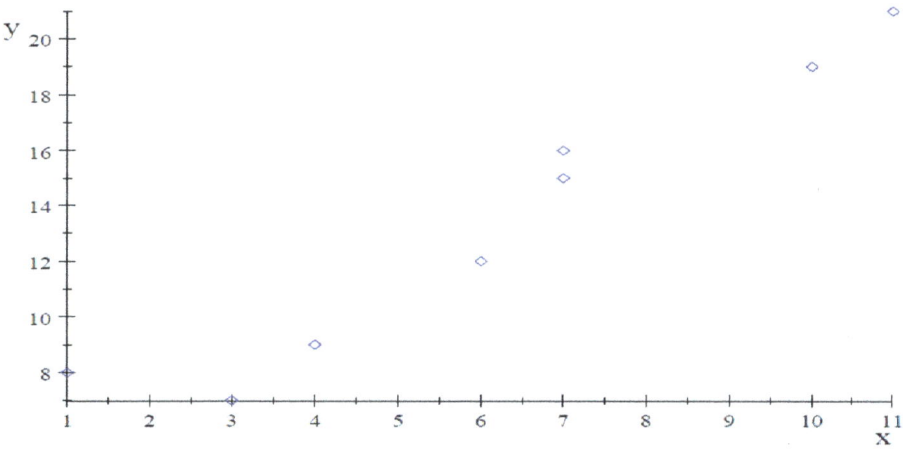

3.1.1 Probability of a Continuous Random Variable

Probability Density Function (pdf). A probability density function is a mathematical function $f(x)$ (abbreviated as pdf) if the following conditions hold:

1. **Positive definiteness.** The function f(x) needs to be non-negative (i.e., f(x) ≥ 0). This condition means that the graph of the function needs to be either on or above the input axis because all outcomes need to be either zero or positive probabilistic outcomes.
2. **Function total.** The total area between the input axis and the function needs to be 1 (or 100%).

Cumulative Distribution Function (cdf). A cumulative distribution function (or cdf) F(X) is the accumulated total of probabilities attained by using the pdf f for a given continuous random variable X. Hence

$$F(X) = P(X \leq x) = f(X \leq x)$$

Probability of a continuous random variable. If f(x) is a continuous distribution, then the probability of x to be between two values of a and b (i.e., likelihood of x to be on the interval [a,b]) can be calculated by using the formula

$$P(a \leq x \leq b) = \int_a^b f(x)dx$$

which is a number between 0 and 1 that represents the area between f(x) and the input axis. We need to calculate the Integral of the pdf f(x) to be able to calculate the total area between the function f(x) and the input axis.

3.1.2 Probability for Discrete Random Variable

Probability Mass Function (pmf). If a random variable X has a discrete random variable nature, then the function representing this random variable (just like pdf for a continuous random variable) is called probability mass function. For instance, if we want to calculate the probability of X to be equal to a specific value a then the probability of X = a is calculated by using the pmf. If f(x) is this pmf then f(a) = P(X = a).

Example 3.2 The following are examples of pmf for different applications:

- **Threat response rate.** Suppose we have a defense mechanism that identifies and defends against threats automatically. If we define the random variable X to be the number of threats responded by the specific system, then calculation of finding the probability of defending against c number of threats would be f(c) = P(X = c) which requires the use of pmf.
- **Inventory management.** Suppose there are several security entities used by a specific organization and the inventory represents these available entities for defending against threats. If X is the random variable representing the successful defenses against threats

by the inventory entities, then finding the probability of a specific entity c to defend successfully would be f(c) = P(X = c) which requires the use of pmf.

Cumulative Mass Function (cmf). It is possible to calculate aggregated total of probability mass function due to its discrete nature that gives the cumulative mass function; hence, cmf is the accumulated total of the probabilities of a discrete random variable X that is calculated by adding up the probabilities up to a certain value of the discrete random variable X. The formula for cmf when $X \leq k$ requires the use of pmf as follows:

$$P(X \leq k) = f(1) + f(2) + \ldots + f(k) (i.e. P(X \leq k) = \sum_{i=1}^{k} P(X = i)$$

Example 3.3 Suppose the goal is to identify the probability of 3 threats' occurrences based on a pre-determined pmf f(x). This probability can be calculated as follows:

$$P(X \leq 3) = P(X = 1) + P(X = 2) + P(X = 3) = f(1) + f(2) + f(3)$$

3.2 Basic Measurement Tools for Six Sigma by Using Basic Statistics

The following basic statistics information forms the fundamentals of the statistics used in Six Sigma projects:

- **Mean (μ).** The mean values of a density function is commonly considered as average of the density function. It is also called expected value, and it is the balance point of the graph with respect to the x-axis geometrically.
- **Range**. The range of a data set is the difference between the smallest and largest data values in a set of data.
- **Mode**. The data value that appears the most frequent among the data values.
- **Median**. After ordering a given set of data values from the smallest to the largest (or the largest to the smallest), the median data value is the data value that is in the middle of this ordered set.
- **Standard Deviation (σ).** Distributions can have the same meaning but completely different spreads about the center. Standard deviation measures how closely the values of the distribution cluster about its mean. If the most values of the input variable are close to the mean value, then the standard deviation is small. If the input values are widely scattered about the mean value, then the standard deviation is large.
- **Variance ($σ^2$).** Variance is used for measuring how much the data varies from the mean.

We will be using the terms mean and standard deviation terms frequently.

- **Central Limit Theorem.** Suppose $X = (X_1, X_2, X_3 ...)$ is a sequence of samples obtained from the underlying distribution. We assume X to be the independent, identically distributed, real-valued random variables with common pdf f, mean μ; and variance s^2. Let

$$a_n = \frac{1}{n}(X_1 + X_2 + \cdots + X_n) = \frac{1}{n} \sum_{i=1}^{n} X_i.$$

and

$$z_n = \frac{a_n - \mu}{\frac{s}{\sqrt{n}}} (standard\ or\ z-score)$$

Then the distribution of the standard score z_n is approximately normally distributed for large enough n.

Central limit theorem indicates that the average of a large number of independent observations from the same population has a distribution that is approximately normally distributed.

In statistics it makes sense to draw a sample from a large population to determine the characteristics of the population from the sample data. Normal distribution can be particularly useful because it is easy to determine the mean value, standard deviation, and ranges of probabilities by using the corresponding normal distribution formulas. For instance, random samples can be collected from an industrial process to determine the stability of the overall process.

A normal distribution can be particularly useful because of the following predetermined standard deviation calculations:

1. Approximately 68.26% of the data falls within 1 standard deviation from the average of the data.
2. Approximately 95.46% of the data falls within 2 standard deviations from the average of the data.
3. Approximately 99.73% of the data falls within 3 standard deviations from the average of the data.

Another theorem that is useful in practice is the law of large numbers:

i. **Law of Large Numbers**

Using the notation introduced in Central Limit Theorem, Law of Large numbers indicate that a number sequence a_n approach to μ as n approaches infinity.

3.3 Discrete Distributions

A discrete random variable is an event that has only discrete possible outcomes.

Example 3.4 Airplane engine production is an example of a discrete random variable. There are only two possible outcomes of this event: Defective and non-defective.

If the event is a discrete random variable, then the probability formula is

$$Probability = \frac{Number\ of\ successes}{Total\ number\ of\ trials}$$

Example 3.5 Suppose defective airplane engine production is the discrete random variable. Let

- p: Probability of a discrete event x to happen.
- x: Defective engine production.
- k: Number of defective items.
- m: Number of non-defective items.

It is possible to calculate the percentage of defective items in the overall production by calculating the simple formula

$$p = \frac{k}{k+m}$$

The probability of getting a non-defective item is

$$1 - p = 1 - \frac{k}{k+m} = \frac{m}{k+m}$$

In this chapter we will cover the following discrete distributions:

- Data driven discrete distribution.
- Discrete normal distribution
- Binomial distribution.
- Poisson distribution.
- Geometric distribution.
- Hypergeometric distribution.

3.3.1 Data Driven Discrete Distribution

A given data set in discrete form naturally represents a discrete distribution. The statistical formulas corresponding to n number of data points $x_1, x_2, \ldots, x_n$ are the following:

- **Mean or Average Value (μ):**

$$\mu = \frac{x_1 + x_2 + \ldots + x_n}{n}$$

Excel Formula: = Average(data)

- **Range of values (R):**

$$R = x_{max} - x_{min}$$

Excel Formula: = MAX(data) − MIN(data)

where

x_{max}: The maximum value in the data set.
x_{min}: The minimum value in the data set.

- **Mode.** The data value that appears the most frequent among the data values $x_1, x_2, \ldots, x_n$.

Excel Formula: = MODE(data)

- **Median.** Manually, after sorting $x_1, x_2, \ldots, x_n$ from the smallest to the largest (or the largest to the smallest), the median data value is the data value that is in the middle of this ordered set.

Excel Formula: = MEDIAN(data)

- **Standard Deviation (σ).** Standard deviation is a measure of identifying how much the data is varying from the standard (where standard is the average). The standard

3.3 Discrete Distributions

deviation formula for a collected data set is the following:

$$\sigma = \sqrt{\frac{(x_1 - \mu)^2 + (x_2 - \mu)^2 + \cdots + (x_n - \mu)^2}{n - 1}}$$

Excel Formula: = STDV.S(data) (when sample is chosen)
= STDV.P(data) (when population is chosen).

- **Variance (σ^2)**. Variance is simply the square of the standard deviation; it helps to measure/quantify the variation in the data set.

$$\sigma^2 = \frac{(x_1 - \mu)^2 + (x_2 - \mu)^2 + \cdots + (x_n - \mu)^2}{n - 1}$$

Excel Formula: = VAR.S(data) (when sample is chosen)
= VAR.P(data) (when population is chosen)

Example 3.6 Suppose a cybersecurity professional collects the following data representing the number of threats and breaches during a specific month:

Data	A	B	C	D
1	2	4	6	8
2	32	25	12	12

In this data set, row 1 represents the number of breaches while row 2 represents the number of threats detected. For instance, cell A1 represents 2 breaches in the system during the first week of the month while cell A2 represents 32 detected threats occurred during this time frame.

In this case the data set values for the number of breaches are A2 = 32, B2 = 25, C2 = 12, and D2 = 12. We can calculate the statistical values as follows:

a. **Mean (μ)**.

$$\mu = \frac{32 + 25 + 12 + 12}{4} = \frac{81}{4}$$
$$= 20.25 \text{ threats occured on average per week during the month}$$

Excel Formula: = AVERAGE(A2:D2)

Hence, we can interpret this response as follows:
   There is an average of 20.25 threats occurred during a specific month of breach observations when threat data is collected every week during a specific month of a cybersecurity professional.

Another way to calculate the average is by using per day. Given that we have 30 days of observations, the average can be calculated as follows:

$$\mu = \frac{32 + 25 + 12 + 12}{30} = \frac{81}{30}$$
$$= 2.7 \text{ threats occurred per day}$$

Hence, we can interpret this response as follows:
   There is an average of 2.7 threats occurring each day of the specific month based on the data collected by a cybersecurity professional.

b. **Range.**

$$R = x_{max} - x_{min} = 32 - 12 = 20$$

Excel Formula: = MAX(A2:D2) − MIN(A2:D2)

c. **Mode.** The data value that appears the most frequently among the data values 32, 25, 12, 12 is 12 therefore the mode of the data set is 12. The Excel formula.

Excel Formula: = MODE(A2:D2) entered in a blank cell would help to find the mode easily

d. **Median.** After ordering 32, 25, 12, 12 from the smallest to the largest 12, 12, 25, 32, the data value in the middle can be either chosen as 12 or 25 since there is no middle value. In this case we can choose 25 which is the closest value to the mean. The following Excel formula entered in a blank cell would help to find the median easily.

Excel Formula: = MEDIAN(A2:D2)

3.3 Discrete Distributions

e. **Standard Deviation (σ).**

$$\sigma = \sqrt{\frac{(x_1 - \mu)^2 + (x_2 - \mu)^2 + \ldots + (x_n - \mu)^2}{n - 1}}$$

$$= \sqrt{\frac{(32 - 20.25)^2 + (25 - 20.25)^2 + (12 - 20.25)^2 + (12 - 20.25)^2}{4 - 1}}$$

$$= 9.9457 \text{ number of threats}$$

Excel Formula: $= \text{STDV.S(A2:D2)}$

f. **Variance (σ^2).**

$$\sigma^2 = (9.9457)^2 = 98.917$$

Excel Formula: $= \text{VAR.S(A2:D2)}$; VAR.S is used sample while VAR.P is used for population.

3.3.2 Discrete Normal Distribution

The distribution has the shape of a bell when the distribution has a discrete probability density function.

Example 3.7 Suppose a cybersecurity professional works on the safety of a network system and understands it better. This required reporting the number of breaches that occurred every year in which case the data can be displayed as a histogram. The following data is collected for the weekly breach attempts in the system.

Data points
9, 12, 14
20, 18, 20, 20
22, 25, 25, 25, 26, 26, 28, 26, 28, 25, 28, 28
29, 29,35, 32, 34, 33, 35, 30, 30, 33, 31, 31, 31, 30
36, 36, 40, 43, 40, 37, 39, 41, 42, 43
45, 48, 48
58

These numbers can be more meaningful if we restructure the data in a distribution format. We split the data into 7 equal weighted groups by calculating

$$Range = \frac{58-9}{7} = 7$$

This allows us to place the numbers in compartments (because otherwise we wouldn't be able to put it in the form in the figures below.) It can be easily calculated that the average of the data is 30.614 and the standard deviation is 9.832. The seven data ranges 9–15, 16–22... form the seven groups of the discrete data. It is important to note here that the continuous formation of such a data set would be very different from the way it appears in the below distributions. One standard deviation range from the average to both left and right sides of the data covers the range (20.782, 40.446). The following image with red marks shows the 1-standard deviation marking range on both left and right sides of the average. This is one way to visualize the statistics on discrete data. It is easy to calculate the 2- and 3-standard deviations from the average.

In this above-mentioned form of regrouping, we can view the distribution of the data as follows:

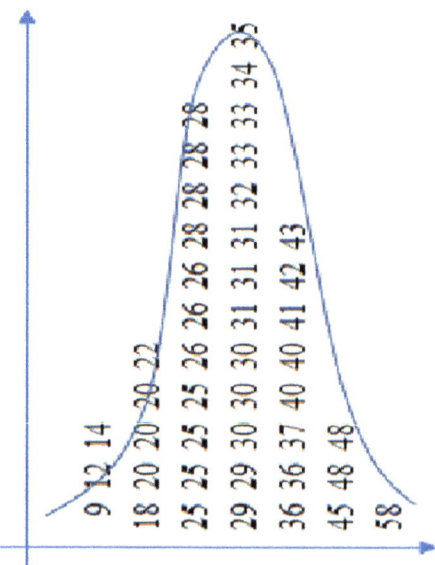

Another way to present this data is by using Excel as follows:

3.3 Discrete Distributions

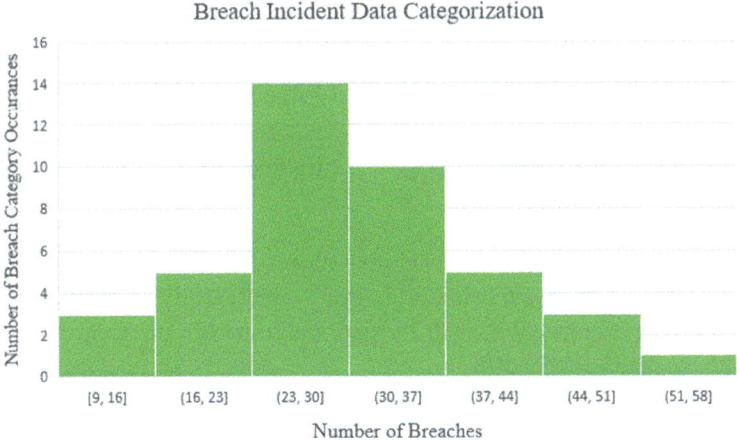

3.3.3 Data with Binomial Distribution Nature

Data that can be attained by sampling with replacement may respect a Binomial Distribution as one of the distributions; This distribution relies on Bernoulli trials that are used in experimenting that should satisfy the following conditions:

- There are n number of trials in the experiment.
- The outcome of the trial needs to be either a success or a failure.
- The change in the trial does not impact the probability of success what should be a constant number during any of the trials.
- All trials need to be independent.

Given a random variable X, it has a binomial random structure if it consists of n number of independent Bernoulli random variables having p as the parameter. The pmf, mean value, and standard deviation of X corresponding to the Binomial distribution are the following:

$$\text{Pmf.} \quad f(x) = \begin{cases} \frac{n!}{x!(n-x)!} p^x (1-p)^{n-x} & \text{if } x = 0, 1 \ldots, n \\ 0 & \text{otherwise} \end{cases}$$

$$\text{Mean.} \quad \mu = np$$

$$\text{Standard Deviation.} \quad \sigma = \sqrt{np(1-p)}$$

Example 3.8 An automated intrusion detection system is determined to detect 96% of the intrusions during a specific month among the 10 attempts. Given this information, we can calculate the probability of one detect attempt:

$$P(X = 1) = \binom{10}{1} 0.96^1 (1 - 0.96)^{10-1} = 10(0.96)^1 (0.04)^9 = 2.5166 \times 10^{-12}$$

Example 3.9 Real life situations do not always allow a distribution to fit perfectly to a data set as can be seen on the graph below. This graph represents data collected for a specific vulnerability being exploited within 20 attempts that respects a Binomial distribution with n = 20 with the probability of 50%.

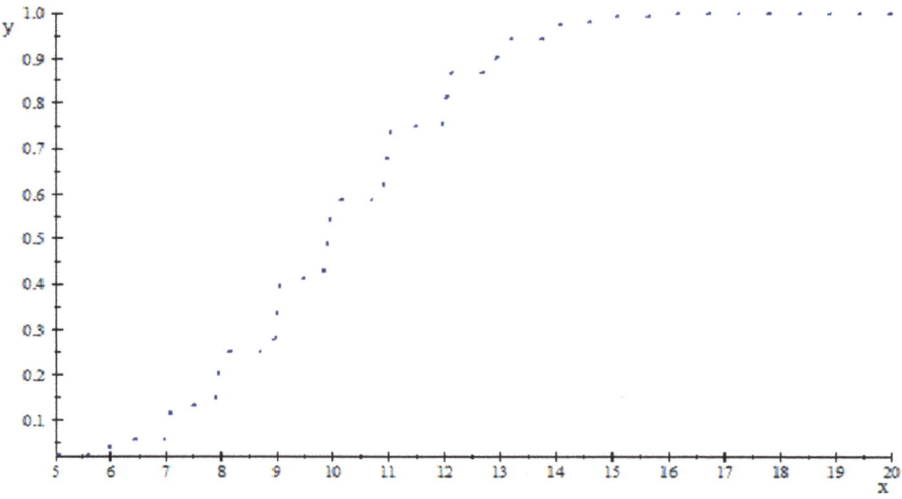

3.3.4 Poisson Distribution

The number of phishing emails received in a day may have a statistical distribution nature with a Poisson distribution fitting to it. As a part of the general definition, Poisson distribution describes the number of events occurring within a fixed interval of time or space. It is used for modeling the number of events that occur when there is a large number of opportunities for the event but there is a small probability that the event occurs on each opportunity. We define the following to introduce the Poisson distribution's formulas:

- X: The number of events/occurrences representing the random variable.
- λ: Average rate of event occurrences.

3.3 Discrete Distributions

The following are the probability mass function, average, and standard deviation of the random variable X that respects a Poisson distribution:

$$\text{Pmf. } f(x) = \begin{cases} \frac{e^{-\lambda}\lambda^x}{x!} & \text{if } x = 0, 1, \ldots \\ 0 & \text{otherwise} \end{cases}$$

$$\text{Mean. } Average = \lambda$$

$$\text{Standard Deviation. } Standard\ Deviation = \sqrt{\lambda}$$

Example 3.10 As mentioned at the beginning of the section; the total number of phishing emails received in a day can be modeled as a Poisson distribution by using the following:

- X: The number of emails arriving at an email account in a day.
- λ: The average number of phishing emails arrivals to email account.

In the case when the probability is small and n is large, Poisson distribution can be used to approximate the Binomial distribution.

3.3.5 Geometric Distribution

Another distribution used for modeling a discrete random variable is Geometric distribution in the case that the number of trials before the first success or failure are modeled by using Bernoulli distribution. Assuming p represents the probability of success (or failure), the probability mass function, mean value, and standard deviation of X corresponding to the Geometric distribution are the following:

$$\text{Pmf. } f(x) = p(1-p)^x \text{ for any non-negative } x \text{ and } 0 \text{ otherwise}$$

$$\text{Mean. } \mu = \frac{1-p}{p}$$

$$\text{Standard Deviation. } \sigma = \frac{\sqrt{1-p}}{p}$$

Example 3.11 Suppose we want to determine the probability of first cyber-thread related failure in a system and this cyber-thread's historical system-based data indicates a geometric distribution nature. If the probability of cyber-thread success is identified to be 0.05 then the corresponding probability of success (or system failure) during the 11th trial by using geometric distribution would be the following:

$$f(11) = 0.05(1 - 0.05)^{11} = 0.05 * 0.95^{11} = 0.02844$$

It is important to confirm that the Geometric distribution is the right distribution for usage prior to using it to make sure that it is a sense making approach for the case to be considered.

3.3.6 Hypergeometric Distribution

Hypergeometric Distribution respects the rule of sampling without replacement with the following used as a part of the associated calculations:

- N. Finite population's size.
- D. Total number of items that are in the group of interest.
- n. Sample size randomly drawn from the population without replacement.
- X. Random variable represents the number of sample items that belong to the class of interest that is drawn without replacement.

The probability mass function of a random variable X respecting Hypergeometric distribution along with the mean and standard deviation values of it are the following:

$$\text{Pmf. } f(x) = \frac{\binom{D}{x}\binom{N-D}{n-x}}{\binom{N}{n}} \text{ if } x = 0, 1 \ldots \min(n, D) \text{ and } 0 \text{ otherwise}$$

where

$$\binom{N}{n} = \frac{N!}{n!(N-n)!}$$

$$\text{Mean. } \mu = \frac{nD}{N}$$

$$\text{Standard Deviation. } \sigma = \sqrt{\frac{nD}{N}\left(1 - \frac{D}{N}\right)\left(\frac{N-n}{N-1}\right)}$$

3.4 Continuous Distributions

In this section we focus on distributions of continuous random variables that rely on events that have continuous flow nature. One basic example is time that is typically used as a continuous random variable as it has a continuous nature. Other continuous random variables that are used in cybersecurity include amount of byte transfer, rating, timestamp etc. Examples of well-known and frequently used distributions include the following:

a. Normal distribution.
b. Uniform distribution.
c. Exponential distribution.
d. Gamma distribution.
e. Weibull distribution.
f. Lognormal distribution.

If f(x) is the pdf of a continuous random variable x, then we can calculate the following:

- **Mean/Average value of a continuous random variable.**

$$Mean(\mu) = \int_{-\infty}^{\infty} xf(x)dx$$

- **Standard Deviation of a continuous random variable.**

$$Standard\ Deviation(\sigma) = \sqrt{\int_{-\infty}^{\infty} (x-\mu)^2 f(x)dx}$$

- **Variance.**

$$Variance(\sigma^2) = \int_{-\infty}^{\infty} (x-\mu)^2 f(x)dx$$

Example 3.12 Suppose the malicious activity handling ability rate of a certain system follows a certain trend that fits to the following formula

$$h(t) = \begin{cases} 6(t-t^2) & if\ 0 \leq t \leq 1 \\ 0 & if\ t > 1 \end{cases}$$

where t is time measured in seconds and h(t) represents the number of malicious activity handling ability rate by the system in hundreds per second. We first need to prove that h(t) is a pdf. The graph of the function is the following graph that clearly shows that it is non-negative (which can also be seen easily through algebraic calculations.)

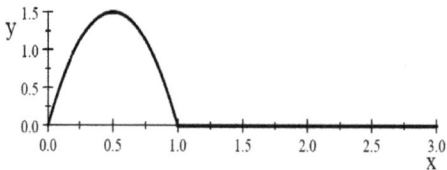

Next, we show that the area between h(t) and the t-axis is 1:

$$\int_{-\infty}^{\infty} h(t)dt = \int_{-\infty}^{0} h(t)dt + \int_{0}^{1} h(t)dt + \int_{1}^{\infty} h(t)dt.$$

$$= 0 + \int_{0}^{1} h(t)dt + 0 = \int_{0}^{1} 6(t - t^2)dt$$

$$= 6\left[\frac{t^2}{2} - \frac{t^3}{3}\right]_{0}^{1} = [3t^2 - 2t^3]_{0}^{1} = 1$$

Therefore h(t) is a pdf.

- Suppose now we want to determine the probability of the system to handle malicious activities within the first quarter of a second. Hence, $0 \le t \le 0.25$ and we need to calculate.

$$P(0 \le t \le 0.25) = \int_{0}^{0.25} 6(t - t^2)dt$$

$$= [3t^2 - 2t^3]_{0}^{0.25} = 3(0.25)^2$$
$$- 2(0.25)^3 = 0.15625$$

This indicates that the probability of the system's ability to handle malicious activities within 0.25 s is 15.625%.

3.4.1 Normal Distribution

The probability density function (pdf) of a normal distribution has a bell shape as shown in the figure below. The pdf corresponding to a normal distribution is the following function given that the mean value μ and the standard deviation σ need to be known for calculations:

3.4 Continuous Distributions

$$f(x) = \frac{1}{\sigma\sqrt{2\pi}} \int_{-\infty}^{x} e^{\frac{(x-\mu)^2}{2\sigma^2}} dx$$

The nature of the pdf of a normal distribution is very complicated therefore it is not easy to calculate normal distribution related information by using mathematical formulas. One easy way to determine the probability by using a normal distribution is by using the following Excel function:

Excel function.   = NORM.DIST(x, mean, standard deviation, TRUE or FALSE)

where the entries listed above represent the following:

- **NORM.DIST** function is the normal distribution function.
- **x** is the value that the distribution of the function needs to be calculated.
- **Standard deviation**. We use the STDV.S() excel function to calculate the standard deviation.
- **TRUE or FALSE**. If we want to calculate the cumulative probability value for x then we need to enter the phrase TRUE otherwise we enter FALSE if we want to find the probability value for x.

A normal pdf is the one that is centered at the mean value at zero and standard deviation at 1; The following is an example of a normal pdf.

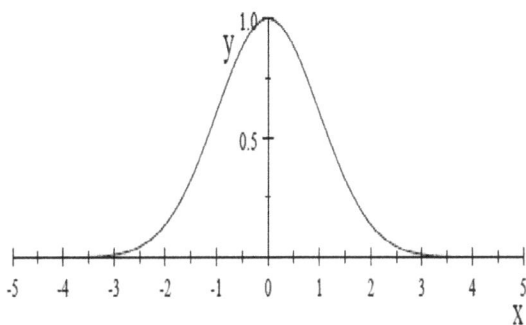

Expecting a perfect match between a data set and a function that fits to it perfectly is unrealistic therefore a pdf typically represents an approximation of the data's structure; hence, the probability calculations that rely on such an approximation function is typically an approximation of the actual value expected to be attained. In such a case, a pdf is used to calculate probability.

Example 3.13 (*Threat Response Analysis*) A particular cyber-threat detection and response system can identify the number of threats and the probability of the threat causing an issue in the system after the review. The use of this system results in normally distributed effectiveness of the threats in the system between zero and 25 threats per day on an average of 15 threats per day and standard deviation of 20 threats detected and responded per day. The number of threats start decaying after 15 threats due to the system's ability to increase the security level. We can calculate the probability of experiencing threats between 5 and 20 per day as follows:

Step 1. Enter input values (just like in the table column) below as 5, 6,...20 in Excel and enter the column **next** to the input the NORM.DIST values. For instance, A2 is the cell corresponding to 5 with the output value B2 = NORM.DIST(A2, 15, 5, FALSE) which is placed in B2 of the table.

# of Threats	Probability
5	0.000514093
6	0.001477283
7	0.003798662
8	0.00874063
9	0.017996989
10	0.033159046
11	0.054670025
12	0.080656908
13	0.106482669
14	0.125794409
15	0.13298076
16	0.125794409
17	0.106482669
18	0.080656908
19	0.054670025
20	0.033159046

Step 2. Highlight all output values that correspond to the normal distribution (i.e., all values in column B) and insert scatter plot to be able to view the following graph.

We can calculate the function = SUM(B2:B17) in Excel that calculates the summation of all probability values corresponding to input values 5–20:

$$\text{SUM}(B2 : B17) = 0.96703$$

3.4 Continuous Distributions

It is important note here that the values of normal distribution range from negative infinity to positive infinity; however, we don't cover these details in this much dept due to its mathematical intensity.

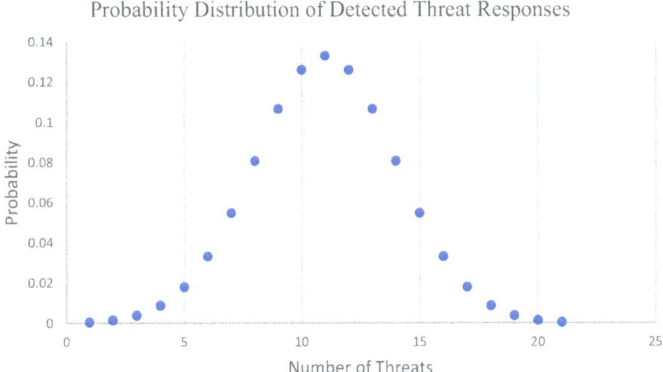

The conditions whether a data set respects a normal distribution structure or not have been studied. Central Limit Theorem states that it is possible to have approximately normally distributed data for a sufficiently large data set under certain conditions. Given a "large enough" data set that respects the studied rules, it is possible to fit a normal distribution to the data. The associated study tests are called normality tests used for identifying the nature of the data from a normal distribution perspective. We benefit from identifying the normal distribution structure of data because it eases the calculations by using formulas and well-known normal distribution results. Software packages such as Minitab have normality tests that can be used.

3.4.2 Uniform Distribution

A continuous random variable that has a uniform nature has a uniform distribution. Such data set that has unified nature of the continuous random variable. A uniform distribution is the most basic continuous distribution. The "uniformity" of the distribution is due to the unchanging probability of the random variable within a range. The general formula used for uniform distribution is the following:

$$f(x) = \begin{cases} \frac{1}{d-c} & c \leq x \leq d \\ 0 & otherwise \end{cases}$$

For instance, if c = 5, b = 25 then the for the random variable x to be between 5 and 25 is 5% and this probability makes the pdf a horizontal straight line as shown in the pdf figure below.

$$f(x) = \begin{cases} \frac{1}{25-5} & \text{if } 5 \leq x \leq 25 \\ 0 & \text{if otherwise} \end{cases}$$

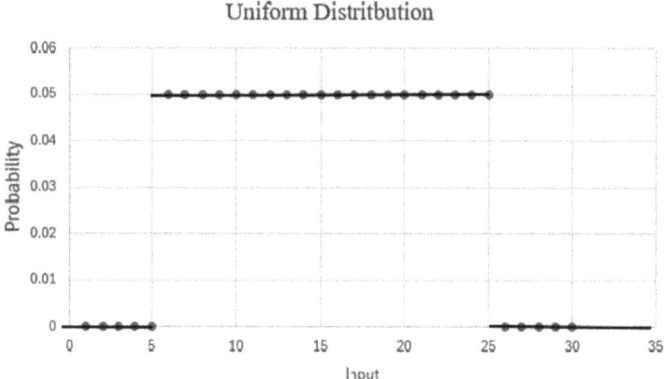

Uniform Distritbution

Example 3.14 The number of transactions processed in a system varies between 50 and 100 network access requests per second with each request having the same chance of occurrence.

The pdf for this statement has uniform distribution nature:

$$f(x) = \begin{cases} \frac{1}{100-50} & \text{if } 50 \leq x \leq 100 \\ 0 & \text{if otherwise} \end{cases} = \begin{cases} \frac{1}{50} & \text{if } 50 \leq x \leq 100 \\ 0 & \text{if otherwise} \end{cases}$$

Hence, the probability of network access to the system is 2% for the users. If we want to determine the specific probability such as the likelihood of access anywhere between 20 and 30 per second, then we use the following formula:

$$\int_{-\infty}^{\infty} f(x)dt = \int_{20}^{30} \frac{1}{50} dx = \left[\frac{x}{50}\right]_{20}^{30} = \frac{30}{50} - \frac{20}{50} = \frac{10}{50} = 20\%$$

Therefore, that probability of network access requests to be anywhere between 20 and 30 per second is 20%.

- **Mean (μ).** The average value of uniform distribution can be calculated by using the following formula:

$$\mu = \int_{-\infty}^{\infty} xf(x)dt = \int_{50}^{100} \frac{x}{50} dx = \left[\frac{x^2}{100}\right]_{50}^{100} = \frac{100^2}{100} - \frac{50^2}{100} = 75 \text{ requests per second}$$

Hence, there is an average of 75 network requests per second on this network.

3.4 Continuous Distributions

- **Standard Deviation (σ).** We can also easily calculate the standard deviation value as follows:

$$\sigma = \sqrt{\int_{-\infty}^{\infty} (x-\mu)^2 f(x)dt} = \sqrt{\int_{50}^{100} \frac{(x-75)^2}{50} dx}$$

$$= \sqrt{\left[\frac{(x-75)^3}{150}\right]_{50}^{100}} = \sqrt{\frac{625}{3}} = 14.43$$

Hence, there are 14.43 occurring requests per second as the deviation from the standard for the specific network.

3.4.3 Exponential Distribution

In real life you may have heard of the expression "exponential increase in quantities" to indicate a dramatic increase in the instance numbers. The mathematical correspondence to exponential function has the following features:

- Only positive or zero outcomes.
- Very large or very small outcomes to occur.
- The value of the largest possible outcome cannot be known.

The pdf of the exponential distribution is the following function:

$$f(x) = ae^{-ax} \text{ if } 0 \leq x \text{ and } 0 \text{ otherwise}$$

where $a = \frac{1}{\mu}$ with μ representing the average. The graphs of the exponential function corresponding when $\mu = 1, \ldots, 5$ are displayed below.

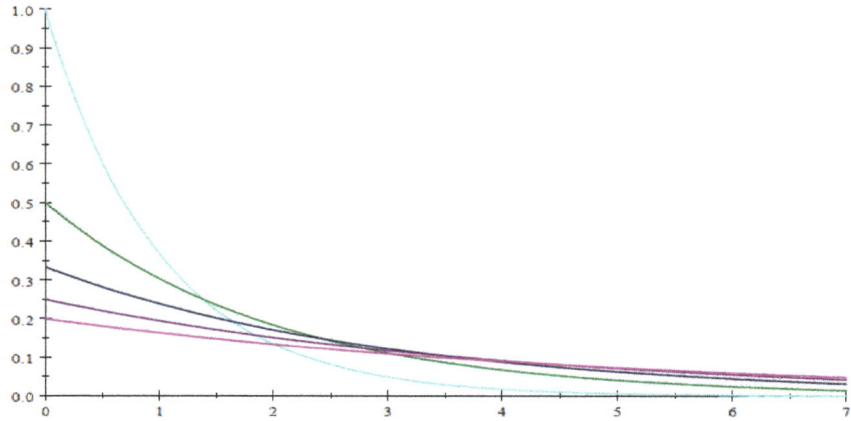

Example 3.15 (*Incident Time to Close*) We use the following information shared in Fig. 9 and design an exponential distribution.

- **Step 1**. Convert data as needed to perform calculations.

  1. Convert all times to integers.
  2. Set identify as time 0 to chart minutes between stages.
  3. Show full recovery in days from identify to resolution.

- **Step 2**. Create an algorithm to chart the data; calculate the differences in between in each stage.
- **Step 3**. Use graphics to show trends in underlying risk. These charts can be used to see if staff can contain in time to avoid damage and fully recover in time to avoid similar attacks.

This data, in combination with other risk attributes, can help determine the probability of a successful attack. The following table contains the associated data. We can calculate the full recovery by using this data.

	Identify	Analyze	Mitigate	Contain	Eradicate	Post-mortem
Network traffic anomaly	0	32	221	97	185	1962
Sensitive data alert	0	5	8	56	3	1412
End point security desktop alert	0	10	8	9	1	117
Server security vulnerability alert	0	5	40	20	0	42

(continued)

3.4 Continuous Distributions

(continued)

	Identify	Analyze	Mitigate	Contain	Eradicate	Post-mortem
Unsecure configuration warning	0	9	39	85	127	630
Email block filter	0	43	23	38	168	1212
OS syslog warning	0	14	20	113	800	494
Detected network anomaly	0	10	30	7	3882	1292
OS process monitor alert	0	8	126	206	629	4982

The following figure is the graph of the pdf:

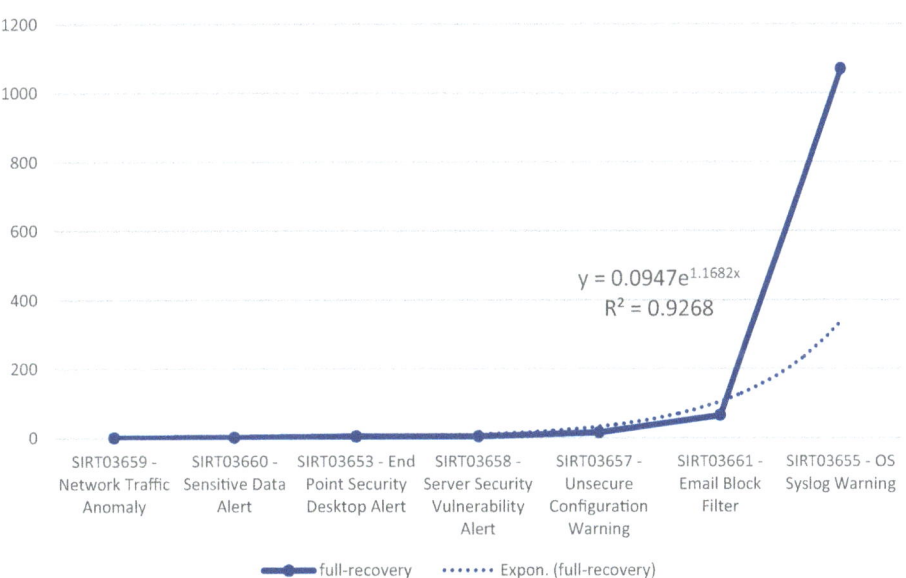

The pdf of the distribution is the following function:

$$f(x) = 0.0947 e^{1.1682x}$$

This function has an R^2 value of 93.95% which is a high value indicating a strong accuracy of the fitting model. This model can help to calculate probabilities associated with time to full recovery. In addition, it provides an insight to the structure of how full recovery is structured.

Example 3.16 (*Security Tool Coverage*) The following is an example of an exponential distribution to be defined in the next section that outlines the tool coverage and the associated performance of the security tool's coverage from a probabilistic perspective. The input axis represents the security tool used with the output axis representing the probability of the coverage of the tool. The detailed information on this example will be explained later in the Define phase. As can be seen on this example, the exponential distribution (green dashed line) appears as a good model with an R^2 accuracy value of 94.01%. We should also note that this distribution may not always be the one that fits to the period data therefore other similar periods of data information need to be analyzed to be confident about the proper nature of this distribution fit in general.

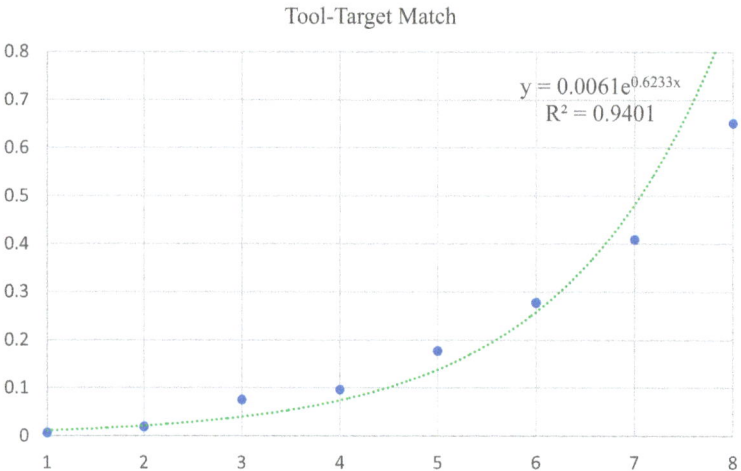

Chapter 3 Exercises

Exercise 3.1 Find and explain the application of a real-life cybersecurity related statistical discrete distribution. What is the nature of data used for such an application and what type of model is obtained from this application?

Exercise 3.2 Find and explain the application of a real-life cybersecurity related statistical continuous distribution. What is the nature of data used for such an application and what type of model is obtained from this application?

Exercise 3.3 Give a cybersecurity application example in real-life that has both a discreet and a continuous random variable simultaneously. What is the nature of such an application and what are the variables used?

Bibliography

1. Montgomery, D. C. (2017). *Design and analysis of experiments.* Wiley.
2. Bickel, P., Diggle, P., Fienberg, S., & Krickeberg, K. (2003). *Springer series in statistics.*
3. Tokgöz, E. (2024). *Quality and lean Six Sigma applications for industrial engineers.* Springer Nature. https://link.springer.com/book/9783031557392
4. Tokgöz, E. (2024). *Quality and lean Six Sigma for engineering technicians, synthesis lectures on engineering, science, and technology.* Springer, 978-3-031-44033-5. https://link.springer.com/book/9783031440328
5. Tokgöz, E. (2025, February) Artificial bee colony optimization techniques' utilization for intrusion detection systems' analysis. In *4th IEEE international conference on AI in cybersecurity (ICAIC) proceedings.* https://ieeexplore.ieee.org/stamp/stamp.jsp?tp=&arnumber=10848880

Let's Define

4

The main goal of the Six Sigma methodology is to incorporate statistics to process improvement for business excellence. The aim is to see the "big picture" of the operations in place and connect data available for analysis to target outcomes and have a holistic solution to identified problems by convening experts in the corresponding fields. In cybersecurity, there are common questions that are an interest in cyberthreat defense systems and continuous improvement is essential due to the need for revisiting these issues. Continuous improvement for resolving these issues will lead to more effective verification and validation for cybersecurity functions. At the highest level, the cybersecurity big picture involves integrating the five NIST functions in such a manner as to achieve the Target Profile. The first step in that process is identifying the Target. NIST guidance for arriving at the Target Profile is to step back and see the big picture from the perspective of Fig. 4.1.

Figure 4.1 demonstrates the "Target" profile that is supposed to be derived from the Risk Assessment. Defining a problem starts with the information attained from threat measurements. Business' ability to withstand the identified threats with the associated risk tolerance of the environment and environmental attributes that drive defense requirements, and the associated resource allocation are essential to ensure that cybersecurity protection measures are adequately provisioned to maintain risk at an acceptable level.

The formation of a six-sigma team depends on the project's scope, and executives' and managers' essential view of the project. Personnel choices for such a project are decided either internally or externally (by a Six Sigma professional or a consultation company.) Potential member choices for a project can include the following:

- Cybersecurity professional

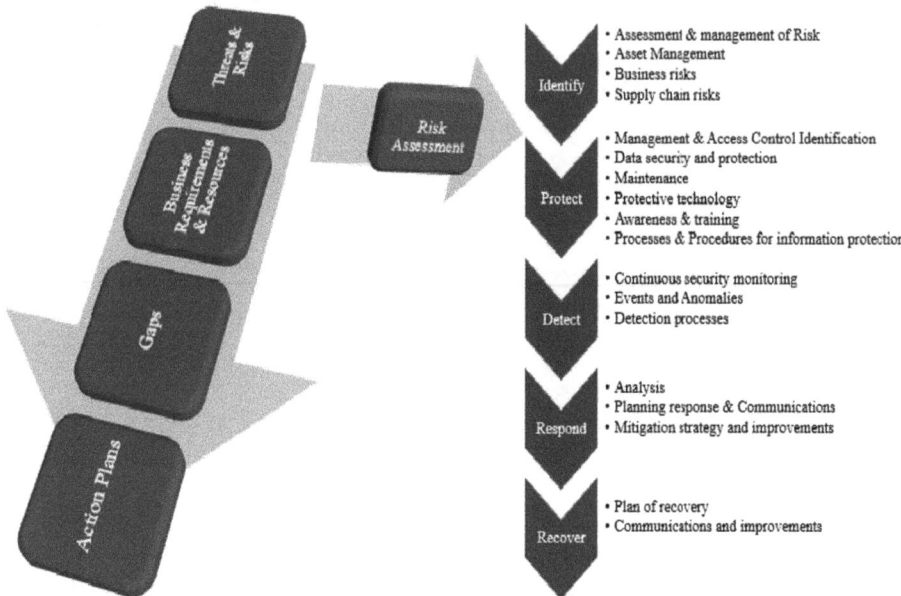

Fig. 4.1 Big picture of risk assessment in cybersecurity applications

- System security officer(s)
- A stakeholder for the project (such as a manager)
- Master Black Belt or Green Belt Six Sigma certified person(s)
- Finance personnel
- Other key personnel taking place in the operations (if applicable)

Example 4.1 (*Incident Time to Close*) Revisiting the *Incident Time to Close* example shared in Example 3.14, we can give meaning to the information shared in the section by defining the problem with an explanation of the information shared.

> There are three different levels of priorities with causes and their effects that result in issues. The operational level of handling Low, Medium, and High priorities is an important factor in handling these issues and finding out what can cause these issues. The time variation of handling issues based on their types and priority levels need to be observed and analyzed to reach operational improvement. Causes and effects need to be investigated to improve issue handling with the time it takes to handle these problems. As a result, the problem definition takes the following form:
>
> What is the impact of issue prioritization on the issues that come up with the time it takes to resolve these issues and how can we standardize the operations to handle these issues?

4 Let's Define

The data set associated with the *Incident Time to Close* example included several variables for consideration that the six-sigma team need to identify which ones to focus on:

- Incident reporting source (e.g., technical operations)
- Type of incident. Malicious, phishing, other, unidentified etc.
- Date.
- Source
- Description
- Cause
- Contact
- Owner
- Priority level
- Contact
- Owner

There are many projects that can be defined based on the above-mentioned information however we will provide only one of them. A typical approach in business is to give priority to the high costing issues to define a six-sigma project's scope and the image indicates that the variational size of the inner radius is the main problem in production.

> Goal. Propose six-sigma solutions to the customer and implement the solution favored by the customer to minimize the incident rate and use of tools while keeping the cost of incidents as low as possible and maintaining (or bettering) the quality of the security level based on CTQ.

From a strategy perspective, depending on the focus of the project, it makes sense to initially drop several data variables from the data set for further analysis that may not be considered as KPIs. This data dropping helps with the simplification of the data set with Contact and Owner dropped from the collected data. One very helpful piece of information in the details of this process would be to include images for each step if possible. Including waste images for the types of waste can also help to visually help the audience to understand the type of defect and locations it may occur.

Example 4.2 (*Security Tool Coverage*)Risk posture in a cybersecurity program requires to protect itself from breaches and safeguard its data that also includes the overall management and strategy related to protecting the enterprise's software and hardware, networks, services, and information. This is an area in cybersecurity in which many six sigma projects can be conducted. There are several key components of risk posture that need to be kept in mind:

- Understanding and defining risk posture for protecting business from cyber-breaches
- Protecting the enterprise from cyber-attacks based on the controls and measures that are identified by the organization

- Effective defense management ability of the enterprise
- Cyber-attack readiness and ability to react to and recover

Given these essential elements of risk posture management, there are key aspects that require establishing a robust risk posture:

- Identifying IT assets and inventories that are included in the protection of systems, applications, devices, data, business processes, and users.
- Defining levels of risks to cybersecurity related assets through vulnerability monitoring and identifying the likelihood of breaches via a range of attack vectors and determining the impact of a breach on a particular asset.
- Identify ways to reduce assets' breach risks that are currently in place (such as firewalls etc.) and document their effectiveness.

The data that is used for the real-life six sigma project to be explained for Security Tool Coverage example contains such data. There is extensive inventory coverage data that cybersecurity firms collect that contains extensive information and it would be easy to define a six-sigma project to further improve the inventory of software used for asset security management.

> The major issue is the mismatch between the tools used and the inventory needed. This mismatch not only causes the company to not match quality expectations but also results in issues related to inventory coverage becoming a bigger issue over time. Therefore, we define the scope of the six-sigma project to be identifying and eliminating tool-inventory mismatch through the analysis of 13944 data points. The resulting outcome is expected to provide the company with information on the need gap of the missing tools for specific operating systems as well as virtual and non-virtual environments. The three-month basis of April, May, and June of 2021 are used as benchmark data for analysis.

Just like in the previous example, there are several KPIs that need to be considered for analysis that need to be measured and some of the information needs to be included. The information excluded are the following:

- IP
- Hostname
- Operating System

The information that is included in the measurements and analysis includes the following:

- Date of incident
- Environment that the incident occurred
- Security tools used

- Virtual or not
- Targeted tool to be used
- Tools available

Example 4.3 (*Vulnerability Patching*) Responding to vulnerabilities is one of the major challenges in cybersecurity that requires careful management with its costs and consequences. We define *Vulnerability Patching* as the attempt to resolve a cybersecurity incident and the time it takes to close such a case upon successful response. For measurability purposes, one can group the threat responses into KPI groups as follows:

- Production Under 30 min
- Production Over 30 min
- Production Over 45 min
- Development Under 30 min
- Development Over 30 min
- Development Over 45 min

The *Vulnerability Patching* example to be used in this book contained data collected during the first 10 months of 2020 with the associated KPIs listed above.

The Define phase requires a careful look at the information on the collected data and expectations of the customer from the customer's perspectives. The expected outcomes of a six-sigma project should be realistic and need to be feasible for completion within the project's expected duration and budget.

The scope of the project and the identification of the essential personnel to work on the project are the primary steps of the project. Once these are determined, the team members that will be working on the project need to be gathered in the "Kick-off" meeting to discuss the project's scope and some of the details. This meeting can also include finalization of the initial important elements of the project. During such a meeting, the following are useful to identify.

- **Project Charter.** A Project Charter contains information on the project specifics that relates to gap closure as it needs to be updated with the progression of the project and the issues that may occur during the progress of the project. This concept will be covered later in this chapter.
- **CTQ.** This component outlines the customer's quality expectations.
- **Project scope.** The project's main scope project is different from the definition of the problem to be stated in the Define phase.
- **DMAIC steps.** Each project is different therefore it is essential to provide an overview of the potential DMAIC steps to be followed based on the initially known scope of the project.

- **End goal.** At the end of the project what is the expected output expected by the customer and what is the end goal of the project?
- **Timeline. The timeline of the project needs to be realistic. Clearly, we ask the question** "What is the timeline for the project?" to be covered in the Project Charter. To meet the expectations of a customer when the timeline is short, it may be essential to adjust/increase the number of six-sigma team member numbers to speed up the project's progress is applicable.
- **Outcome's impact.** An impact analysis that is measurable is always helpful. Hence, the expected impacts of the outcome for the customer and how these expectations should be reflected to the outcome can be outlined.
- **Project needs.** The scope of a project requires fulfillment of certain tasks, and a project has "needs" therefore project needs should be identified based on the project's scope.
- **Resource gap closure.** The completion of a project would have resource requirements for its completion that need to be identified for closing the gap from a resource perspective. Such a gap closure also includes the gap closure for quality, technology, staff needs etc. that depends on the scope and goals of the project.

The definition of the project expectations is the key component of the Define phase as it is the first phase of the DMAIC approach in Six Sigma. Each step to be taken for a project's completion might be well known and implemented in a work place during ordinary and daily works completed; however, this doesn't indicate that ways utilized for handling such tasks are in alignment with customer expectations and the customer may not be willing to pay for all the techniques used for completing these tasks and activities that do not add value to the product or service. Activities that are not adding value (NVA) can be realized only by observing and evaluating the process carefully. The problem to be solved through the Six Sigma project should be stated clearly, and details of the input, process, and output should be documented by using diagrams. Such diagrams can be structured based on the wastes considered. The following steps are the highlights of Define phase:

- It is possible to clearly sketch operations on a paper by using symbols (icons) with the flow of the operations clearly outlined and explained in a diagram.
- Actions taken in the process and the timelines of task completion can be placed on maps.
- It is essential to identify the weaknesses in the processes based on the waste categories.
- Identification of the VA and NVA activities and clearly outlining them on the process map based on the weaknesses determined would be helpful.
- Documenting where, when, how, and by how much there can be improvements made in the process to outline how project's solution may be implemented would be beneficial if there are opinions on the solution of the problem outlined during the Define phase.
- Clearly outlining the locations of potential improvements on the sketched map with the determined improvements by using icons would be useful.

- Determining how much value the potential improvements can add and eliminate NVA to the current operations may be possible to calculate.

The following tools can be used during the Define phase to complete the steps above.

- Value Stream Mapping
- Input-Process-Output diagram
- Supplier-Input-Process-Output-Customer diagram
- Project charter
- Critical To Quality (CTQ) tree

The right selection of the problem and the associated project scope during the Define phase is critical to the success of the project to be able to implement improvements on the most critical needs of operations. This selection usually depends on the leader of the top management or sponsor of the Six Sigma project. Reducing cost is one of the objectives of Business while threat detection and elimination is another scope that relates to the cyber-thread analysis. During the Define phase it is important to characterize the following for the project:

1. Purpose
2. Coverage
3. Objectives
4. Expected outcome
5. Budget
6. Resources
7. Lead time

The diagrams used during the Define phase can be helpful for visual demonstration of the expectations and known scopes of the project in addition to the progression of the project after the Define phase.

4.1 IPO Diagram

Input-Process-Output steps can be outlined by using an IPO diagram as a basic diagram that helps to visually represent the work progress in simple terms. For instance, in cyber-security, the input can be the materials and software used for certain operations that relate to the scope of the Six Sigma project while the processes utilized include the processing to attain the outputs based on the inputs used. It is possible to increase the complexity of the IPO diagram that can be extended to incorporate more technical engineering work if needed. In this diagram customers and suppliers are not considered as a part of the work in progress.

- I: Input
- P: Process
- O: Output

Figure 4.2 is an example of an IPO diagram with inputs used for making bread and the output attained that is attained as the bread. The workflow in this diagram is progressing from Input to Output while identification of CTQ conditions depends on the reversing the workflow steps from Output to Input as the feedback would be attained from the customer based on the Output's quality. For instance, when you bake bread for the first time, the customer can provide feedback based on the tested bread and comment on what needs to be improved. This would mean modification of the process first and identification of the input based on the modified process steps.

Sketching of an IPO diagram can be accomplished by following the steps below.

- **Step 1:** Select a process for improvement.
- **Step 2:** Improvement is a success only when the output that is attained as a product or service obtained from the process is considered to be a success by the consumer/customer. Therefore, it is important to determine the expected quality conditions of the output product or service that will be valued by the customer that is known as the Critical to Quality conditions (CTQ).

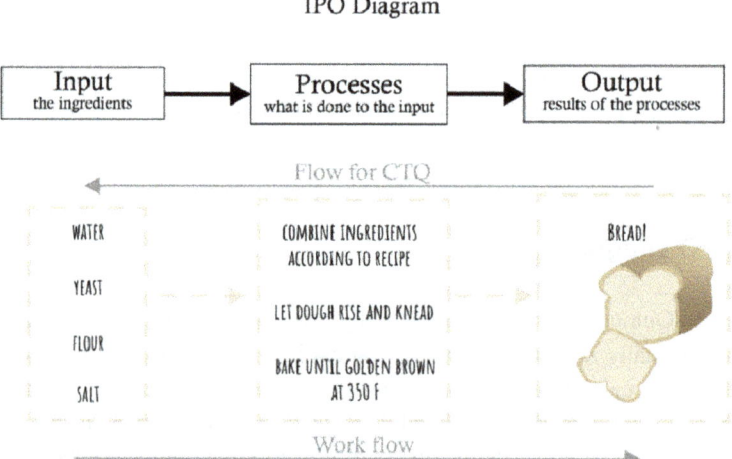

Fig. 4.2 IPO diagram steps that can be followed for preparing bread with the CTQ flow to be progressing from output to input and workflow progression would be from input to output

4.1 IPO Diagram

- **Step 3:** Identify the required inputs using the CTQ conditions determined in Step 2 and sketch an IPO diagram with details outlined for each one of I, P, and O.

The flow of information is initiated by the output requirements and workflow is initiated with the input. A SIPOC diagram is a diagram that needs to be structured by incorporating the active involvement of the Customer and Supplier by incorporating them into the IPO diagram that will be explained in the next section.

Figure 4.3 is an example of a CTQ tree that integrates the voice of the customer. Such integration helps to identify customer-driven decision making wherever it applies. In addition, VoC helps to identify the "Drivers" that take place in between the CTQ and the needs of the customer to outline improvement opportunities. It is a good strategy for identifying the improvement opportunities based on the CTQ information attained from the customer and possible root causes for issues that occur based on the "needs" analysis.

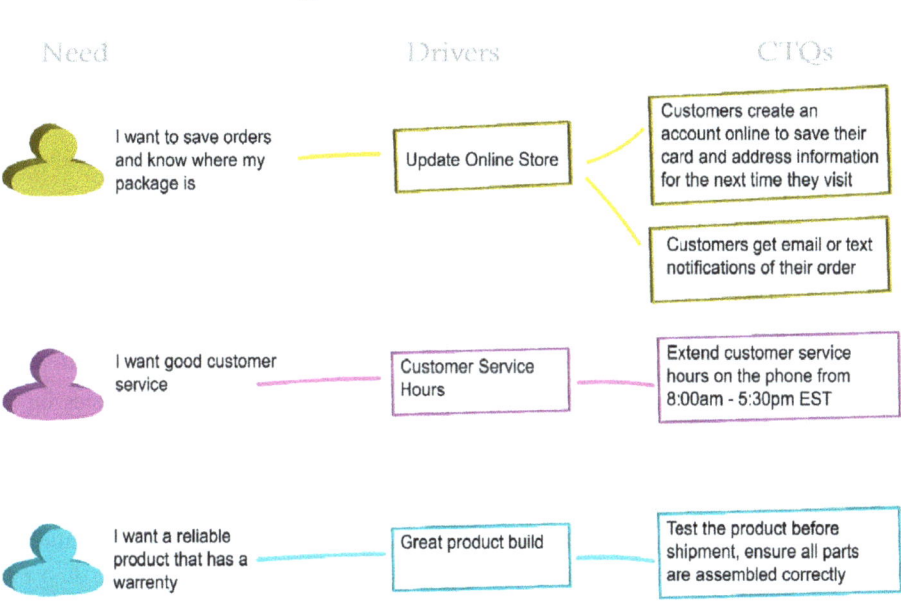

Fig. 4.3 CTQ tree structure that is structured by using the voice of the customer

4.2 SIPOC Diagram

A Supplier-Process-Input-Output-Process (SIPOC) diagram is constructed just like an IPO diagram with the incorporation of the Supplier and the Customer in the IPO, therefore the following are the elements of a SIPOC diagram.

- S: Supplier
- I: Input
- P: Process
- O: Output
- C: Customer

In cybersecurity, a supplier can be a vendor of software used for thread analysis, it could be the physical drives purchased, the cloud service provider etc. The following steps create a cycle for CTQ conditions:

- **Step 1:** The information received from the customer identifies the CTQ conditions to identify the Output expected for the project.
- **Step 2:** The Output of the project determines the structure of the processes and the associated management to attain the expected conditions.
- **Step 3:** The specific process steps determined for the project lead to the identification of the conditions for the input to satisfy.
- **Step 4:** The project inputs determined the specifications received from the identified processes to determine the supplier.
- **Step 5:** The identification of a supplier is based on the specifications of the project and the workflow initiates the product.
- **Step 6:** New CTQ conditions may be added by the customer to the project with newer information that can result in repeating the steps outlined above and updates on the project.

Figure 4.4 is a demonstration of the cyclic SIPOC steps that starts with the customer initially. However, this cycle may not have this cyclic behavior all the time after the first time, meaning it may not necessarily follow the order of S-I-P-O-C after first time around. For example, after the steps of SIPOC are applied, the customer's modification to the input may be requested rather than process therefore Input step would need to be visited first to modify the process. Application of SIPOC steps end once the customer is satisfied with the CTQ conditions based on the output attained. As outlined in Fig. 4.5, it is possible to make a connection between the SIPOC diagram and a flow diagram that illustrates the flow of the work from the beginning to the end. Such a mapping can help to identify the details of SIPOC and where the process mapping makes a connection for the project.

4.2 SIPOC Diagram

Fig. 4.4 SIPOC cycle with the beginning of the cycle initiating with the Customer

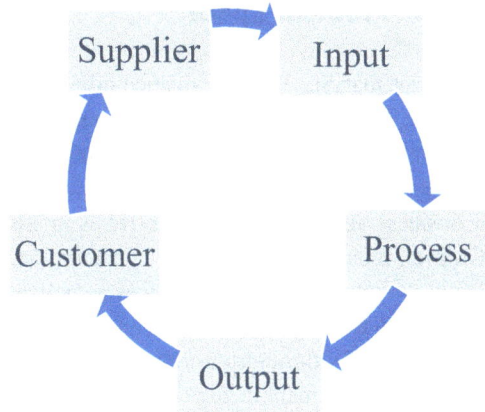

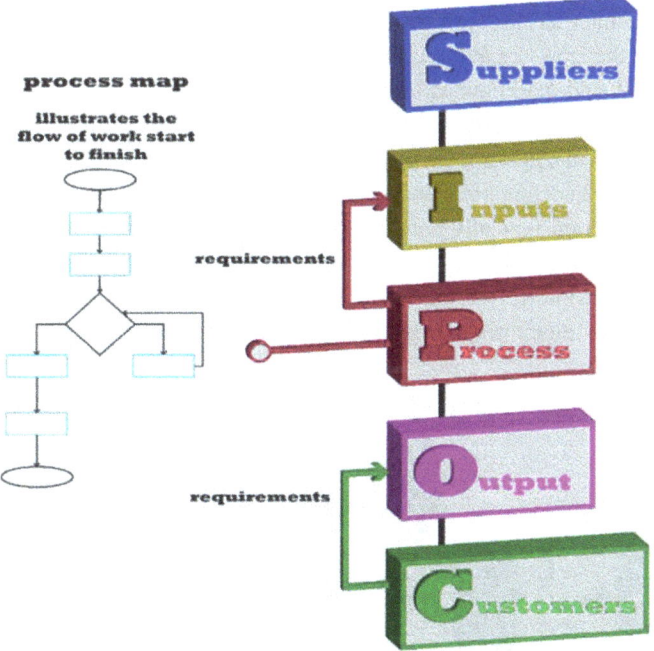

Fig. 4.5 SIPOC diagram in relation to the process mapping with the flow

4.3 Problem Space

One way that Six Sigma guidance advises to think about this kind of problem space is to think in terms of the organization's production workflow in the context of its environment. Figure 4.13 is an example of a "SIPOC" diagram. As implied by the arrows connecting the objects in the diagrams, a SIPOC diagram documents the workflow from a Business Supply Chain perspective as Inputs to internal Process, which in turn produces Outputs, intended for Customer consumption. Although Cybersecurity professionals typically do not think of their services as deliverables external to the organization in which they serve, Six Sigma focuses strongly on the mindset of customer satisfaction and problem definitions through customer analysis. Cybersecurity problems that appear to have little impact on manufactured products used by business customers may nevertheless be analyzed through a Six Sigma lens by focusing on the overall business performance that includes business continuity, regulatory and contractual compliance, and public relations goals that have impact on business capability to serve its customers. Definition of cybersecurity problems through the lens of Six Sigma will necessarily frame the problem as constrained by internal and external forces that pressure the organization's capability to accomplish the goals of its cybersecurity program. SIPOC and IPO have the same information flow and workflow as shown in Fig. 4.6:

Another way to think about the big picture from a cybersecurity perspective is from the perspective of managing a cybersecurity project as its own internal process, which has unique inputs and outputs, constraints, and opportunities. For example, inputs to a cybersecurity program typically come from business management and regulators. These are used to establish a company policy on what information is important to protect as well as the responsibilities of individuals to implement the policy and monitor policy compliance. While monitoring, protection gaps may be identified that demonstrate the

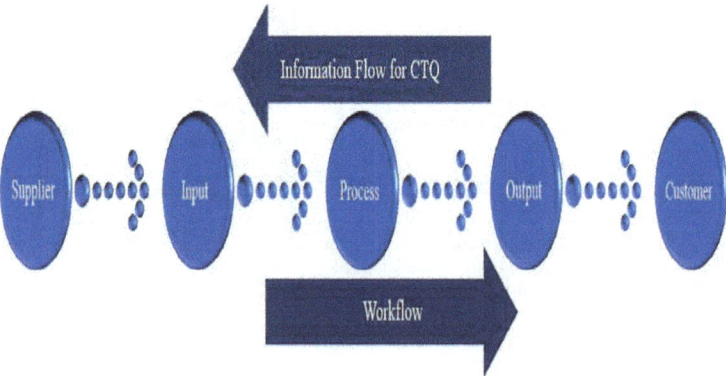

Fig. 4.6 SIPOC Diagram with the information flow direction for attaining CTQ and workflow

need for process improvement, and new strategies are devised that contribute to business requirements. Both the SIPOC diagram and the Cybersecurity management diagram may be productively used to identify the problem space and help specify strategic targets for process improvement that, if accomplished, would make the business more resilient against cyber threats. To illustrate the application of Six Sigma methodology to cybersecurity incident response, consider that cybersecurity monitoring tools extend broadly throughout business technology infrastructure. Even the most well defined and executed process must change as the technological landscape changes around it.

4.4 Lean Six Sigma

Enterprises that adopt Six Sigma methods typically engage in process improvement projects as problems are identified. Technology problems routinely surface during incident resolution. In enterprises with significant digital assets to protect where such a problem even peripherally involves a cybersecurity function, that problem is typically prioritized for process improvement. Many organizations that adopted Six Sigma find that the same DMAIC methodology, if the measure and analysis phase include incorporate cost, serves to increase efficiency as well as performance in terms of desired outcomes. A special feature of six sigma is its ability to produce measurable outcomes that demonstrate the difference between the existing security operations and potentially improved operations. Such results are attained upon the completion of Lean and Six Sigma techniques that incorporate statistical significance. Table 4.1 compares Lean Six Sigma to Kaizen event approaches.

In this Section we will focus on improvement of cybersecurity processes via the Lean Six Sigma methodology, where problems within processes are examined in the context of customers, inputs, process, outputs, and suppliers.

4.4.1 Customers

Cybersecurity defense processes never exist in a vacuum. It is critical to their success to understand exactly what they are meant to defend. When management appoints a Chief Information Security Officer ("CISO"), it is typically assumed that the scope of the CISOs defense responsibilities include the entire enterprise. However, it is often the case that a new CISO, meeting with the CIOs who support various organizations within the enterprise, may find one or two operations that have historically eluded centralized enterprise control processes and the expectation that this exclusion will apply to the CISO's information security program.

Even where it is clear that the entire enterprise is the beneficiary of a CEO-mandated cybersecurity program, it is often not clear to a CISO exactly whom within the enterprise

Table 4.1 Comparison of Kaizen event and Lean Six Sigma project properties

Properties	Kaizen Events	Lean Six Sigma
1. Statistical analysis	Relatively simple	Extensive that can include statistical distributions, queueing theory etc
2. Training	Only training for the specific event	Professional (green and black belt training)
3. Duration	Long period	Short period
	(Example. 3–5 Months)	(Varies; possibly 1–3 weeks)
4. Teams	Multidisciplinary	Multidisciplinary
5. Team member professions	Sponsor & Specific discipline-trained professionals (such as engineers, quality specialists etc.)	Sponsor (champion) & Six Sigma trained professionals
6. Main goals	Applying basic principles of Lean Methodology:	Statistical analysis for
	1. Eliminating Waste	1. Quality improvement by defect reduction
	2. Cycle time shortening	2. Yield improvement
	3. Improvement of workflow	3. Making decisions based on the collected data
7. Involvement	Full time	Part time (can also be full time)

is the voice of the customer. Some CISOs work closely with the development community to support secure software deployment and may treat the development managers as customers. Others work closely with the risk management teams who oversee their work and may treat them as customers. However, from the perspective of the business process, the customer is always the customer who is served by the business. Where an information security program is part of the business process, they will have the same customers as the business process. So, it is best to avoid the mirage of an internal customer that can serve as proxy for the business customer and simply for the CISO to acknowledge that their function is most effective is cybersecurity serves the needs of the customers of the business. It is the role of the CISO not just to align security operations with a business process, but to integrate with business process so that the customer is served by cybersecurity and cybersecurity is an aspect of the quality of the goods and services offered to the customer by the business. This way, there is little room for confusion on whether security measures are appropriate or not. The same decision process can be utilized to provide a quality product provides a secure one.

4.4.2 Inputs

As human computer interaction evolves in complexity, humans develop tolerance for inconvenience in pursuit of automation. A good example of this is the proliferation of multi-factor authentication and security questions. It is increasingly common for a service website such as a bank, insurance company, or email provider to interrupt a normal user login process because of some security algorithm that has determined that the user's behavior is out of the ordinary. For example, the user may be accessing the service from a hotel or café when they typically always log in from home. In such situations, the login will be interrupted. The user will be asked to identify objects in blurry pictures that only a human eye could pick out. They will be asked where they went to grade school, or a numeric code will be texted to their phone. Their use of the service will have been delayed for 3 min. Yet they will tolerate the interruption because they are accustomed to making trade-offs in time for security.

The example notwithstanding, there are some limits to user tolerance for security measures, especially if they seem easy to bypass. For example, if a service is free and it takes less time to enroll in the service than to discern captchas and answer security questions, they may just enroll under a different name to avoid the security inconvenience. Therefore, a principle of "user acceptability" is a common cybersecurity requirement. The service cannot just implement cybersecurity measures that it believes will maintain its own goals, say of identifying the same user across time to sell records of user behavior to advertisers, it also must support the cybersecurity trade-offs made by its user community. The level of customer participation in the protection mechanism must be acceptable to the customer as a fair trade-off that is inextricably connected to the benefits of the service.

4.4.3 Process

Over the course of the previous decade, the process-oriented framework for cybersecurity defense has been defined by several national and international standards bodies, a leader in which has been the US National Institute for Standards and Technology (NIST). In 2014 a Framework for Improving Critical Infrastructure Cybersecurity is published by NIST as a customizable process that combines input from multiple cybersecurity professionals rather than a set of prescriptive requirements. Its target was to form a critical infrastructure, and it has been used by organizations to form a good security posture. It includes five functions, and the corresponding categories and subcategories of activities recommended to be included in an organization's functional implementation of cybersecurity measures. In this framework, four tiers of increasing degree of rigor and sophistication are defined as a part of cybersecurity risk management practices: Partial, Informed, Repeatable, and

Adaptive. They are meant to help an organization determine the extent to which cybersecurity risk management is informed by business needs. The NIST definitions of functions that are common across enterprise cybersecurity processes are defined as follows:

- **Identify.** Develop an organizational understanding to manage cybersecurity risk to systems, people, assets, data, and capabilities.
- **Protect.** Develop and implement appropriate safety to ensure delivery of critical services.
- **Detect.** Develop and implement appropriate activities to identify the occurrence of a cybersecurity event.
- **Respond.** Develop and implement appropriate activities to act regarding a detected cybersecurity incident.
- **Recover.** Develop and implement appropriate activities to maintain resilience plans and restore capabilities or services that were impaired due to a cybersecurity incident.

Each function includes activities that are combined into an organization's cybersecurity operational process. The activities overlap and support each other to allow simultaneous execution and continuous improvement. As discussed above, for them to be meaningful, they must be tightly integrated into the business process that they support. In the example business process above, they would be incorporated as follows:

4.4.4 Outputs

- The activities in the Identify function are foundational for effective use of the Framework. They are methods of understanding the business context, the resources that support critical functions, and the related cybersecurity risks enable an organization to focus and prioritize its efforts, consistent with its risk management strategy and business needs.
- The Protect function supports the ability to limit or contain the impact of a potential cybersecurity event.
- The Detect function enables timely discovery of cybersecurity events.
- The Respond function supports the ability to contain the impact of a potential cybersecurity incident.
- The Recover function supports timely recovery to normal operations to reduce the impact from cybersecurity incidents.
- Although each function has specific outcomes that provide value to the business, the overall outcome of any cybersecurity process is the control over information dissemination, maintenance of data integrity, and the availability of automated information processing.

Example. The following are the outcomes of the associated functions:

- **Identify function.** Asset Management; Business Environment; Governance; Risk Assessment; and Risk Management Strategy.
- **Protect function.** Identity Management and Access Control; Awareness and Training; Data Security; Information Protection Processes and Procedures; Maintenance; and Protective Technology.
- **Detect function.** Anomalies and Events; Security Continuous Monitoring; and Detection Processes.
- **Respond function.** Response Planning; Communications; Analysis; Mitigation; and Improvements.
- **Recover function.** Recovery Planning; Improvements; and Communications

4.4.5 Suppliers

Today's cybersecurity space has critical importance in vendors' technology infrastructure, platforms, and services. When compared to industries such as aerospace and health care providers, suppliers in cybersecurity do not as tightly engage with the user community as much as these industries. Cybersecurity supplied products typically claim to target and solve the vast majority of CISO concerns as their customers in which case CISO requirements are depending on the specific operational needs of the organization. Due to the specificity of the cybersecurity operations, CISOs' expectations for suppliers to match the associated needs require additional efforts unless the security is custom designed for the specific system. Given the complexity of systems, a CISO's ability to develop all the technologies needed to automate the full range of functions described is almost impossible. The choice of supplier and the associated products typically targets to support the operational processes required to execute those functions. It is possible to optimize the number of products used for fulfilling the functions through the right choices of suppliers and products. These products form the inventory of products.

4.5 Cybersecurity Process

As noted previously, at a high level, cybersecurity processes of different organizations have enough similarities, however, at a more detailed level, even within an industry, the technology that supports the business process varies widely. Hence, even though NIST has high level process definitions that serve as a guidance for cybersecurity process development, every organization must have a self-developed framework for ensuring that cybersecurity process provide appropriate coverage from a business perspective.

The following is the list of NIST's framework for process illustration of cybersecurity functions:

- Identify
- Protect
- Detect
- Respond
- Recover

NIST emphasizes the importance of cybersecurity framework (CSF) to be adaptable to the industry sector, a variety of technologies, and technology lifecycle phases, therefore the design of a CSF can be compliant with reducing cybersecurity risk composed of its three parts:

- Framework Core
- Framework Profile
- Framework Implementation Tiers

Customization of the risk prevention techniques with the desired outcomes are expected to be structured and published in a catalog. The use of best practices for risk assessment and prevention, and the associated lessons learned by the stakeholders from incidents with updated best practices are natural to occur during continuous improvement of CSF design. Figure 4.7 outlines a systematic approach to the management of NIST's framework for process illustration of cybersecurity functions.

It is possible to sketch cybersecurity operations on paper by using universal icons and design a table/diagram that includes flow of the operations by answering the following information:

- Process Phase?
- Owner?
- Purpose?
  - Identify catalog scope of business cyber dependency and protection requirements
  - Identify compromised information and technology assets
  - Implement controls to reduce cyber risk to an acceptable level
  - Restore business functionality after damaging cyber events
  - Specify alerts on compromise, investigate, and contain damage
- Coverage area(s)?
- Technology use? Management of...
  - Compromised Business Application
  - Configuration
  - Identity and Access

4.5 Cybersecurity Process

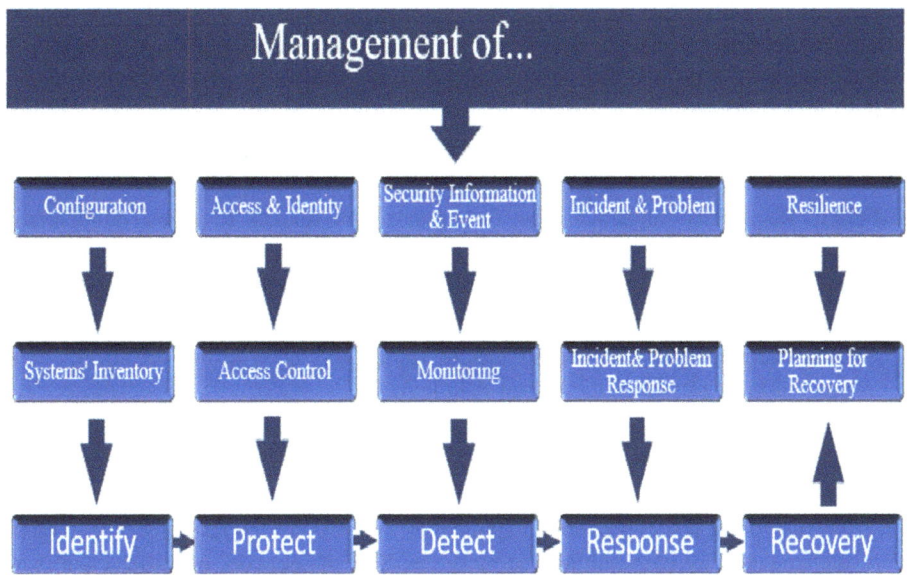

Fig. 4.7 Management of cybersecurity operations at three different levels

- Incident and Problem
- Security Info and Event
- Procedure(s) used?
- Outcome(s)? Expected results on…
  - Business Resiliency
  - Information Security Standards
  - Risk Management Framework
  - Security Operations
  - Technology Forensics

Just like VSM designed for business operations, VSM icons can be used for outlining cybersecurity operations by sketching them on a paper by using universal symbols (icons) with the operational flow and logistics of the operations clearly explained on a diagram.

Establishing a connection between organizational policies, procedures, and standards with the expected cybersecurity functions to protect, defend, respond, and recover phases is recommended by NIST for supporting cybersecurity activities. Accumulation of data attained from cybersecurity tools help to structure the backbone of cybersecurity infrastructure of an organization. Given a process, layered policies, standards, and procedures of an organization uses to communicate its decisions on how the process will be executed within the enterprise is established through cybersecurity control hierarchy as follows:

- **Process.** Anything that takes place between the output from a workflow perspective to support a given outcome. The design of a process needs to be driven by the roles, responsibilities, size, and nature of the organization due to the Cybersecurity Governance requirements.

    Identify, Protect, Detect, Respond, and Recover are examples of functions that are for sub-processes taking place in the operations.
- **Policy.** Regulatory objectives for managing risks related to data confidentiality, integrity, and availability form the basis of a methodology to comply with management needs. Management mandates are formulated as policies for the security program.

    Job functions within the organization need to identify the operating system and server access rights and the number of such rights need to be minimized for the tasks at hand to reduce the risk of data leakage.
- **Standard.** Technical norms that are directives within the organization for used technical configurations that comply with policy either at organization or department level.

    The use of an ID on production application systems can be restricted to emergency maintenance activities authorized by the head of the network operations center and facilitated via the CyberSecurity Vault system.
- **Procedure.** Step-by-step instructions for describing process steps.

    For example, step by step instructions are provided to employees to instruct technology configurations and ensure that standards are followed and train new personnel.
- **Guidance.** The staff needs to be guided by the right methods of system access and communication channels along with other matters that require guidance. An example of such guidance is the passcode selection of the employees.

An example of a template containing procedural steps that can be followed by a staff to respond to an incident is given in the following table. The information in such a template can be updated as the procedure to follow changes when needed.

Owner	CISO
Name	Incident Response
Purpose	Contain damage
Organizational Coverage	Department level
Technology Domain	Infrastructure
Preconditions	Problem report
Tools Used	Incident and Problem Management System (IPMS)
	Cybersecurity Forensics Toolkit (CFT)
	Security Info and Event Management System (SIEM)
	Raw logs and configurations from system of interest

(continued)

4.6 Document Problems

(continued)	
Work	1. Select highest priority alert identified by (IPMS)
	2. Review log details and interpretation guidance
	....
Measures	Fields in records of procedure execution in an incident and problem management system
Outcome	Successful containment or problem declaration
Time & Success ratio calculations	Minimum, maximum, and average times of completion
	Success ratios for of completion

The Organizational Coverage and Technology Domain needs to be covered by the organizational coverage and technology domain of the procedures must map to and fully cover in the process definition. In this case, an *unplanned work* is a gap identified. Value of the gaps and outcomes are determined based on the cybersecurity risk reduction. The value of cybersecurity to the business depends on the cybersecurity operational process. Risk assessment, project management, training, and reporting are additional planning and supporting processes to the cybersecurity operational processes. Given that customer is the key factor in a Six Sigma project, it is very important to include these processes in a Six Sigma project for ensuring that input and suppliers are well defined, and the expected outputs are attained upon completion of cybersecurity operational processes to provide customer protection and satisfaction. These support processes should be targeted at the same level of analysis as operational processes to ensure that waste is identified and eliminated for operational success.

4.6 Document Problems

Indicators of cybersecurity related problems are often identified as incidents prior to them being declared as successful attacks. Any incorrectly configured control mechanism presents a problem, and any activity that breaches an authorization process presents a problem. Bookkeeping is essential in cybersecurity for documenting lessons learned and using such information for similar applications. As controls are documented, there should be corresponding documentation of what constitutes a control bypass. To enable precise descriptions of control bypass requires precise description not only of a control but of the technology component within which the control is configured. Figure 4.8 shows how controls are documented and how technological components are measured. Multiple data fields populated by using technology components are captured by different data-gathering processes to provide information on a common set of systems attributes. Note that this figure includes data elements that are not specific to the technology itself but incorporate how the technology is managed by the organization; that is, information of the engineer

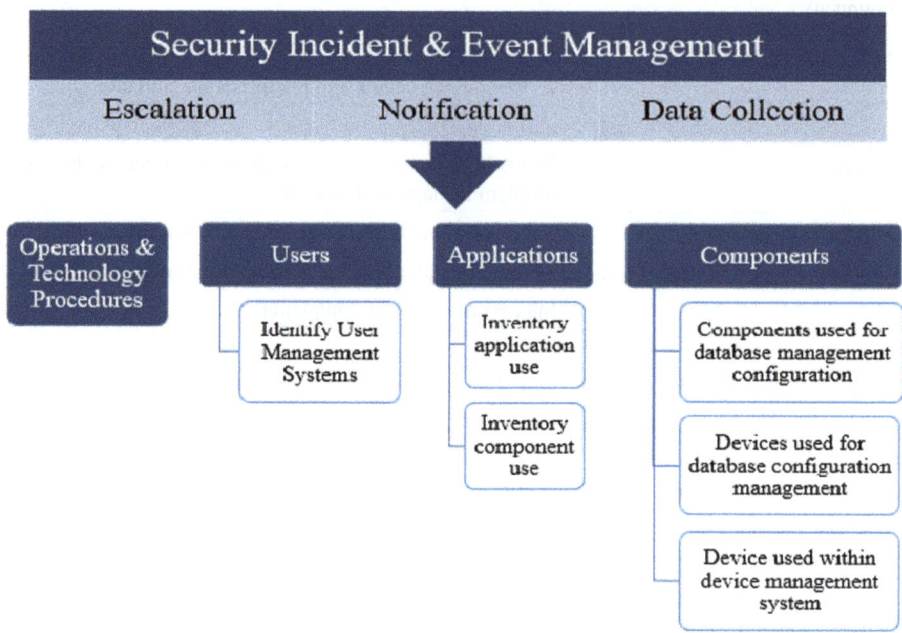

Fig. 4.8 Technology measurement considerations with security incident and event management

or manager's identity who is accountable for the associated correct configuration, and the communications path by which problems with the technology are escalated.

One way to measure is technology differentiation based on different types of technologies used. Technology *controlled* by management is another measurement from a cybersecurity perspective. Effectiveness of different technologies based on data collected can be measured through a thorough analysis to identify each individual technology effectiveness based on its impact to the business. Information corresponding to customer valued technological expectations should be identified during the Define phase. There are measures that provide the basis for metrics and the associated correct and reliable measurements need to be conducted. Attributes of technology control measures include the following:

- **Source.** Demonstrates the data repository from which the measurements are collected.
- **Scope.** Data mapping to a business process, organization, or enterprise.
- **Algorithm**. The technique used for measurement of the data
- **Interval.** The measurement time with specific periods shown if exist.
- **Unit**. Measurements units used.

The cybersecurity technology that underlies a business application is expected to be readily available for ease of access in the cybersecurity excellence center, therefore considerable effort should be made to ensure that such data is available; The contrary would result in wastes that need to be eliminated. Such data forms the inventory used for cybersecurity applications that support business processes. Continuity of Business Plans and Technology Incident Management, and Information Security Management Process are examples of such processes.

4.7　Flow Diagram (Chart)

A flow diagram can represent a process by using simple universal icons and explanatory text that would help to outline the steps of a process to be followed from beginning to the end of the process:

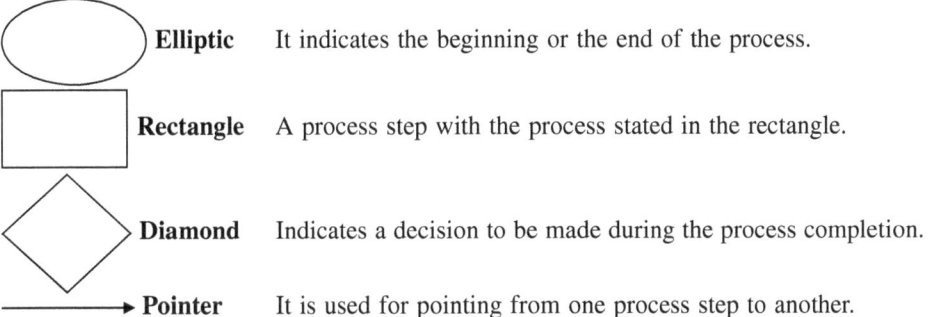

	Elliptic	It indicates the beginning or the end of the process.
	Rectangle	A process step with the process stated in the rectangle.
	Diamond	Indicates a decision to be made during the process completion.
	Pointer	It is used for pointing from one process step to another.

The following steps can be followed by the team members to construct a basic flow diagram:

- Identify the beginning and ending descriptions of the process based on the goals of the project.
- Identify the major steps of the process-flow from top to bottom.
- Identify the key decision-making points in the process.
- Use the flow chart's universal symbols to generate the flow diagram.

A flow diagram is a representation of the necessary steps to be taken from beginning to the end in the general process that represents the summary of the entire process flow with decision making steps included. A well-developed flow diagram can be an effective way of outlining SIPOC diagram and overall process steps to be taken, however flow diagrams need to be developed carefully without skipping important flow steps and the decision-making steps' specifications. Some of the important considerations for successful flow diagram development requires paying attention to the following:

- Meaningful text with short phrases for all the steps taking place in the flow chart
- Right symbol choices would reflect the intended meaning within the flow chart
- Correct step selection with the right number of steps chosen that play a key role in the project
- The right decision-making steps should be selected

An activity flow diagram similar to a flow diagram can be designed with all activities included for the operations taking place around the work to be done. It can be viewed as a larger scope flow mapping that can include additional operations that can impact the project indirectly. Figure 4.9 is an example of such a mapping that can be designed with the operations color coded based on the locations that they are completed.

Example 4.4 Suppose there is a need for handling a vulnerability detected as a part of a Six Sigma project to redesign its method of handling by using the tools used for vulnerability patching as a part of a certain operation. The following flowchart outlines the steps to be taken.

Activity Flow Mapping

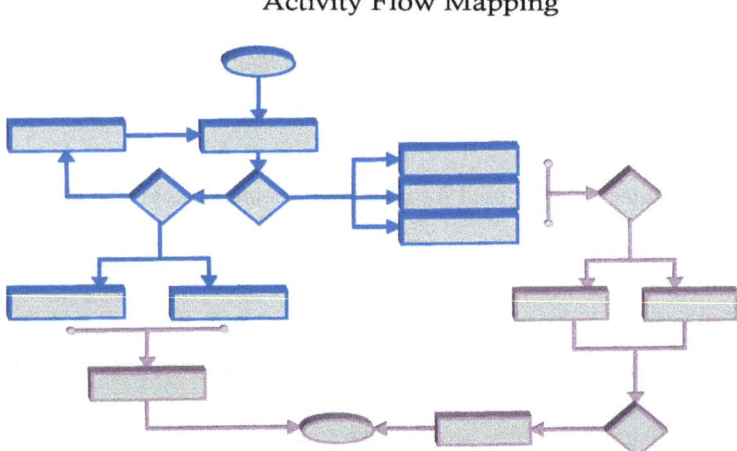

Fig. 4.9 An example of an activity flow mapping template that can outline one or more operations

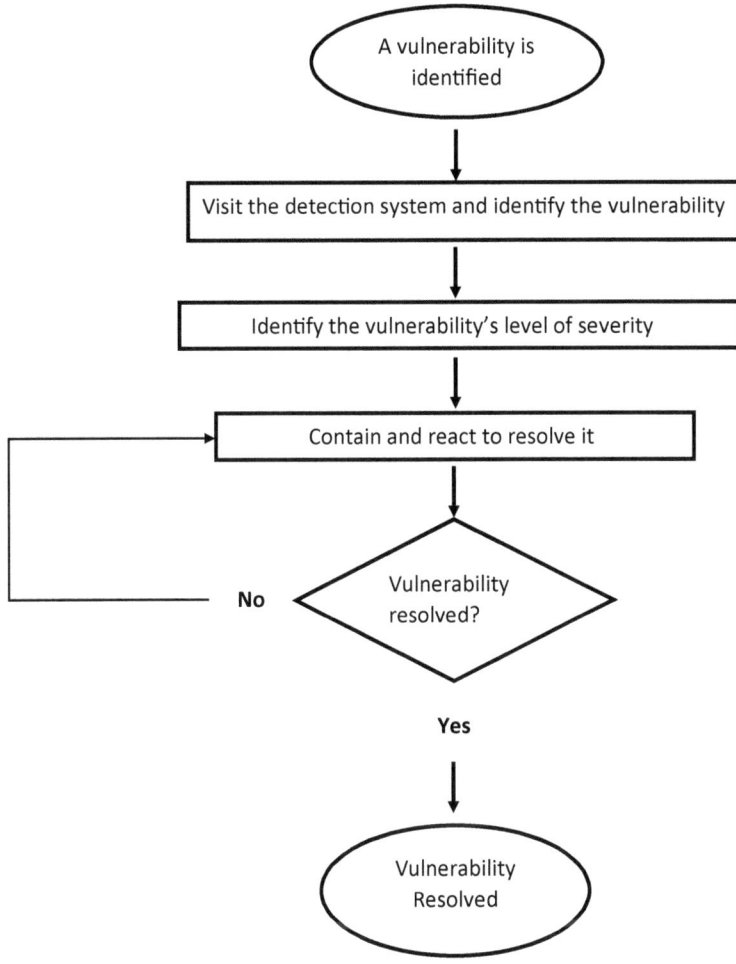

4.8 Business Process Mapping

All steps in a process from the beginning to the end can be visualized by using Business Process Mapping. In order to develop such a map, a detailed look at the production/service from top to bottom is needed. Some of the details of the production/service that needs to be included in a business process mapping include the following details at each step of the process during the production/service:

- Process Initialization
- Duration of sub-processes to be completed at each level
- Process costs at each level per unit time

Fig. 4.10 A flow chart design by grouping VA, BVA, and NVA

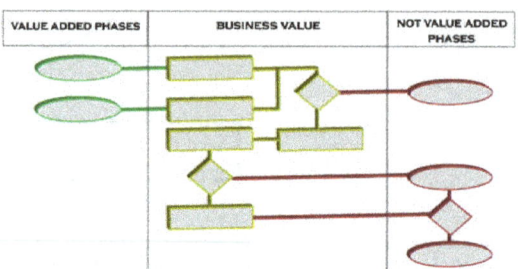

- Entire production's (or service's) total cost
- Overall process duration
- Process outcomes
- Specific unit of the business that deals with the specific sub-process

A process may require several subprocesses to interact and run altogether, therefore a flow chart may have a complex structure for large operations. It is possible to design a flow chart based on the VA, NVA, and BVA values as demonstrated in Fig. 4.10 with the two inputs evaluated to be VA, the three outputs attained to be NVA with the business value added activities outlined in the middle that includes process and decision-making steps.

A more structured diagram of the general process that includes the business units from the beginning to the end is a (Business) process mapping. Such business units may include the initial threat response team, finance, data analytics team etc. In a business process mapping, the processes are placed under the business unit based on the involvement of the unit to complete the steps.

4.9 Functional Flow Diagram

We define a function in a business to be the duty of a business department/unit to accomplish a task. For instance, the cybersecurity unit of a business would be responsible for threat detection and response. In this regard, a functional flow diagram is a flow diagram that is designed based on the functions defined within business. For instance, some of the functional units in an organization are the following:

- Information Technology (IT)
- Research and development (R&D)
- Cybersecurity specialist
- Data science team
- Marketing

4.9 Functional Flow Diagram

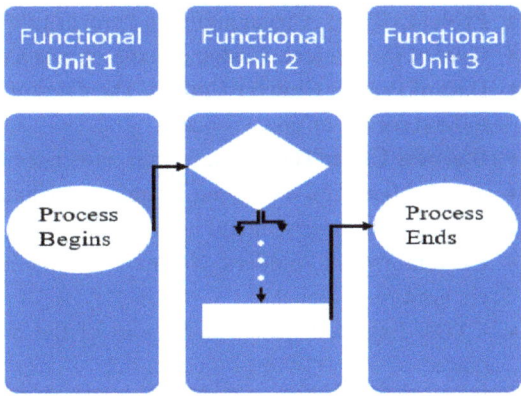

Fig. 4.11 A basic template that can be used for designing functional flow diagrams

- Accounting/finance

The meaning of function depends on the tasks to be completed for a project that may change due to the functional units taking place in the operations and changes in responsibilities. The design of such a functional flow diagram depends on the scope of the project and functional specifications required for a project. Similar to Fig. 4.10, a functional flow diagram can be split into multiple compartments with the flow diagram elements to be placed under their respective functional units as demonstrated in Fig. 4.11.

For instance, if we consider the *Security Tool Coverage* example (Example 4.2) then the units included are network specialists, cybersecurity analyst team, IT, data analysts, operations team, finance, and program manager. Hence, we can categorize the functional units to be categorized into the following:

- Tool inventory elements and coverage management (program manager)
- Analysis of the past threads (cybersecurity analyst team)
- Effectiveness analysis of the inventory on thread responses based on the open and closed cases (data analyst team)
- Finance (costs of inventory software and thread response costs)
- Flow of actions taken for thread response operations and their effectiveness (operations and cybersecurity teams)

4.9.1 Project Charter

Documenting the project with certain important aspects and updating it throughout the project period are important for a successful project and a Project Charter is a form that helps to accomplish it. The following are some of the key elements of a Project Charter:

- **Project's Title:** A meaningful and proper title assigned to the project.
- **Goal of the Project.** Customer expectations should be outlined for the project.
- **Date of Project Charter Update.** The dates of the updates on the Project Charter need to be recorded as updates occur.
- **Business Units.** The business units associated with the project need to be outlined.
- **Team Members.** All team members including the leader, sponsor, stakeholders, mentors, and others should be stated.
- **Customers/Stakeholders.** Names of the internal or external customers and stakeholders need to be stated.
- **CTQs.** Expected quality conditions of the customer/stakeholder need to be written down.
- **Project's Description.** A brief description of the project.
- **Project's Scope.** A detailed explanation of the project based on the goal and the title of the project. The level of the detail is up to the Six Sigma team.
- **Budget and Relevant Requirements.** The costs of the project and relevant requirements need to be explained.
- **Beginning and End of the Project.** Expected beginning and end dates of each one of project's DMAIC phases; Define, Measure, Analyze, Improve, and Control.
- **Improvement Expectations.** A brief outline of the expected improvements upon the completion of the project.

Formal communication among the project team members through updated versions of a Project Charter is necessary for a smooth progress tracking of a Six Sigma project. Such changes are needed as there are updates on the project. For instance, there may be an expected end date required by the customer and this date can change based on the progress of the project unless the customer requires a "must end" date. Noting that the information determined for the Project Charter mainly consists of expected improvements, it is natural if it requires changes. In special cases it may not have to change; For instance, if the terms of the project are required by the customer and if the project must be finished under those conditions with the strict due date in a short period of time, then it may not be reasonable to update a Project Charter. Updates on a Project Charter may occur during any DMAIC stage of the project after its initial development during the Define phase.

4.9.2 Spaghetti Diagram

There is a need for displaying the interaction between interactive components of a system and a Spaghetti Diagram is a useful technique to accomplish it. In a system, there are movements of entities and information flow during task completion and communications. In many systems such actions are repeated many times periodically. A simplistic improvement idea implemented in such systems' movement flow can add big value to the process,

4.9 Functional Flow Diagram

however guessing what might be a potential improvement is not enough; it should also be observed and documented. One of the ways to improve the workflow and operations is through the movement flow of the entities and/or communications with respect to the sources of such communications. Such movements should be sketched in a meaningful way on the service map of the workspace. Such observations should be completed for each process completion cycle and this cycle should be re-observed to make sure the movement flow observed at the first time is proper. If the operations are too big to fit in one spaghetti diagram, then it may be ideal to subgroup operations to design several interrelated Spaghetti Diagrams. If it is possible to record the movements within the system, then such recordings can be used for accomplishing Spaghetti diagram sketching. Such a diagram can be used during the completion of both Kaizen events and Six Sigma projects. For example, possible improvements in the workplace after sketching a spaghetti diagram and making the necessary changes can include the following:

- The response time it takes to communicate between two locations where output coming out of one location is the input to the other location.
- Operational hours of the service provider or the party responsible for the service. For instance, one of the thread detection software may be used for handing certain threads in certain locations within a network; however, after observations made by using a spaghetti diagram, changing the utilization areas of the thread detection software within the network can result in dramatic improvements.
- Physical and/or digital (that can be measured in distance and time) traveling of entities within the network from one place to another to
  - Complete tasks
  - communicate
  - process requests
  - collect data

Figure 4.12 is an example of a spaghetti diagram for cybersecurity operations with 6 different colors representing different groups of locations. Examples of such locations can be the following:

- DNS server
- Machines
- Cloud storage locations
- Servers
- Buildings
- Facilities

The curves can represent the pathways of entities. Examples of entities include risk assessment related countermeasures to cyber-attacks and information flow within service facilities:

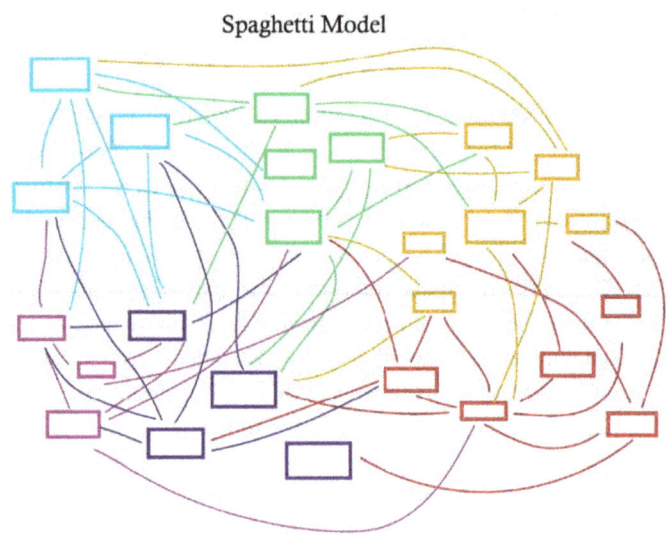

Fig. 4.12 A spaghetti diagram based on the grouping within the work environment

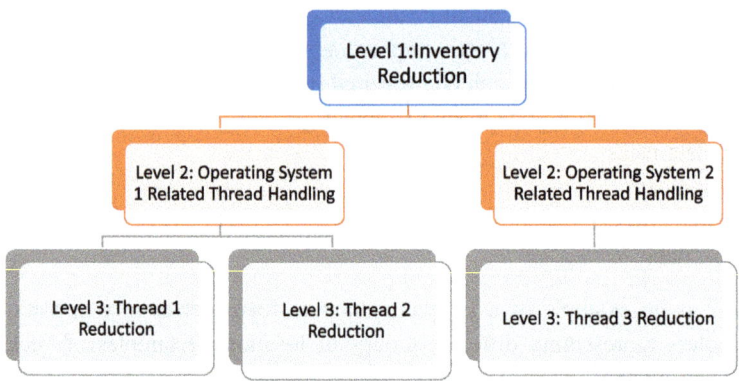

Fig. 4.13 Critical to Quality hierarchy tree for software inventory management

- Employee physical movements
- High volume data flows
- Requests and responses
- Authentication between locations

The color coding is designed to indicate different units where the flow of entities occurs.

For instance, it may be challenging to identify the factors that affect the pathways of traveling entities, and pathways of such entities may not be necessarily a straight pathway that may be traveling, and this causes waste of time and resources. Reallocation of traveling pathways of entities can result in straight pathways between the operating machines and shortening the pathways; Therefore, it is possible to decrease the distance traveled by the entities by making such lengthy pathways shorter. It is important to test the system to understand if the suspected changes are causing the expected improvements or not. Such improvements can yield time and cost savings to the business and reduce the possible stress on the systems.

4.9.3 Hierarchy Tree of Critical to Quality Conditions

The critical components that describe the hierarchy of expected accomplishments to achieve Critical to Quality (CTQ) conditions can be designed as a tree by employing a CTQ hierarchy tree' It describes the critical components of the process that the customer values as quality categories in the product. This tree typically contains specific and general customer requirements depending on the scope of the project. Such requirements are driven by the expected customer/stakeholder outputs from the process that is valued the most by the customer/stakeholder and the CTQ conditions specify the output's quality categories that are determined by the customer/stakeholder. The identification of CTQ categories can be designed by collecting data from customers. The analysis of the collected data can reveal the customer valued outputs and outline how the system works; therefore, the CTQ tree can also be used in the Analysis phase. The top-level customer expectation is the first level of the CTQ tree, and it outlines the most general output expectation of the customer. The following levels are designed to break down the top-level expectations into smaller expectations. The number of CTQ tree's levels indicates the depth of the customer expectations from the output. Several considerations during the design of a CTQ tree can include the following:

- What is the primary step to be taken to start the project by using CTQ?
- What is the top level (i.e. level 1) category of CTQ expectation?
- What is the 2nd level CTQ expectations?
- What is the 3rd level CTQ expectations?
- Continue the fulfillment of levels until the CTQ details are covered.

Example 4.5 (*Security Tool Coverage*). Suppose a part of the Six Sigma project's goal is to focus on improving software inventory covered in Example 4.2 and the CTQ requires to reduce the security tools by identifying the unnecessary tools that are not fully utilized with possible duplicate efforts. Figure 4.12 initiates with the highest level of this goal to be as inventory reduction. The second level can be designed to further specify the effectiveness of the software used for thread handling for each one of the operating systems used by the organization. The third level can furthermore specify the effectiveness of the software used for thread handling by the associated software. We can incorporate numbers into the tree to indicate measurable effectiveness.

One other way of structuring the CTQ hierarchy tree in Fig. 4.13 could be based on the 2nd level consisting of the specific software that is in the inventory. Level 3 can be used for specifying the types of threads handled by using such software. The associated measurable considerations can include the effectiveness percentage of software for handling specific threads by using the historical data, the time it takes to handle threads etc. We can incorporate depth to the tree by including level of vendor effectiveness and other CTQ conditions as required for the project.

4.9.4 Initial Phase of Value Stream Mapping—Defining Operations

A Value Stream Map (VSM) is a very powerful tool as a sketch of project-related operations by using universal icons. It is typically used in many different settings in cybersecurity such as network communications, cloud operations, cyber-thread analysis and other service operations. It makes sense to sketch a VSM in cybersecurity with a meaningful representation assigned to icons.

- All the operations (such as security level assignment, communication method, access level and authentications, data flow, requests for authentication, responses to requests, etc.)
- Locations such as users, Web servers, application servers, financial servers, database servers, etc.
- The amount of time spent during each step.
- VSM should include the lead time (i.e., total time spent reaching output from supplier to the customer), value added, business value added, and non-value-added activities with the corresponding time frames pointed out. The following are some of the universal icons used for structuring a VSM:
- ⓞ Icon demonstrating personnel.

- 🏭 XYZ Icon representing customer and supplier.

- | Weekly Schedule | Icon used for displaying information.

- | STEP1 / NAME OF PROCESS | Icon representing the process Step 1 and its name

4.9 Functional Flow Diagram

- 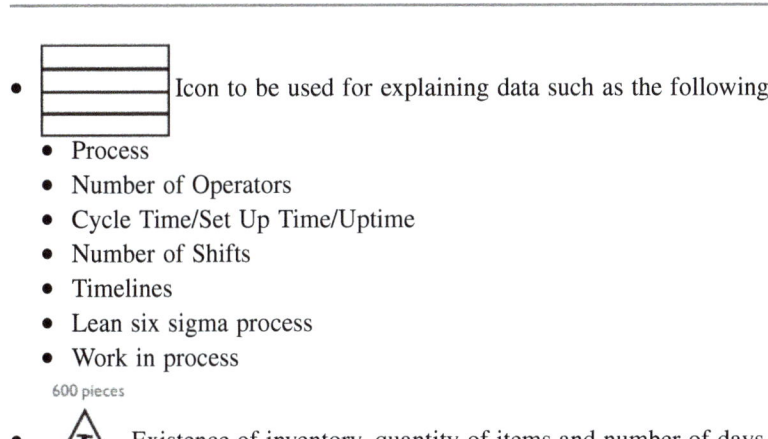 Icon to be used for explaining data such as the following:
  - Process
  - Number of Operators
  - Cycle Time/Set Up Time/Uptime
  - Number of Shifts
  - Timelines
  - Lean six sigma process
  - Work in process

-  Existence of inventory, quantity of items and number of days can be included

- Indicator or an improvement opportunity within the process

- External communication that indicates completed task movement.
- Material movement using the "push logic": Completed task is sent to the next step in the process even if it is not needed in the next step.
- Server

- Computer

- Router

-  Electronic information flow.
-  Manual information flow.
-  Buffer (e.g. 3 days).

- Pull or withdrawal method.

-  First-In-First-Out.

- $\boxed{\text{O X O X}}$ Level load production icon to indicate levelling of production quantity rather than batching.

- 👓 Inspection requirement.

-  Lead time overview by using VA and BVA, and NVA times.

The total of the VA and NVA are entered at the end of the lead time of the entire map.

- ⊔ A Kanban post is used for processes replenishment.

- ▱ Withdrawal Kanban is a method for transferring entities from location to location.

- ▭ Production Kanban is used for upstreaming to produce the entities needed.

- ▽ Signal Kanban signals the need for a new entity production due to reaching the re-order point.

Figure 4.13 is an example demonstration of the VSM by using some of the universal icons outlined above. In this graph, each process step can be the work completed by servers with the inventory representing the storage space and relevant logistics required by the process step. The electronic communication between the process steps and data storage that is asssociated with cloud storage is also outlined for the network. There is truck transportation of the storage units from supplier to the inventory of the facility while there is also physical item transfer from the last step of the process to the customer at the last step of the process. The lead and cycle times can be entered at the bottom of the image based on the measurements collected from the system.

Six Sigma teams typically design VSM by drawing them with a white or black board and edit it as there are updates on it. Another way to design a VSM is by using web-based software (Fig. 4.14).

The second VSM phase is the improvement phase with all identified Six Sigma project improvements ideas sketched on the initial VSM phase. The third and last VSM phase

4.9 Functional Flow Diagram

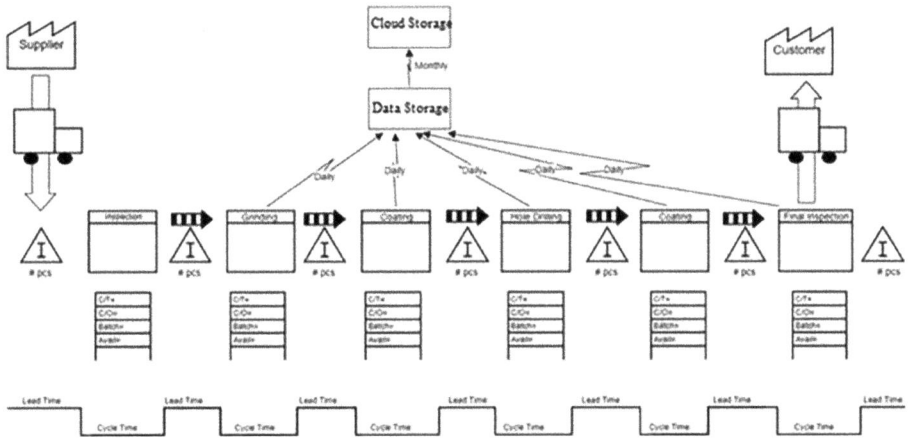

Fig. 4.14 An example of a VSM without specific numerical values outlined

reflects the implemented improvements on the initial phase of VSM that forms the new business strategy that reflects the improved system.

4.9.5 Kano Model

A Kano model is a visual representation of customer satisfaction based on the actions taken by the Six Sigma project team in the improved process. The three levels of customer satisfaction as the following:

- Delighters
- Satisfaction
- Dissatisfaction

Some of the process steps may have higher precedence over the others based on how much the customer values them:

- An improvement in a specific area of the process that the customer was not expecting to see improvements upon the completion of a Six Sigma project is called a **delighter**. Such an accomplishment may occur naturally during the project completion phase that adds value to the project and makes the customer/stakeholder happier since it was not expected.
- Satisfaction is accomplished when the sub-processes that are expected to be improved by the customer are completed in the way that the customer desired. Such tasks have a higher priority for improvement than the others from the customer's perspective;

therefore, they require the highest priority in the list of tasks to be completed from a successful project completion perspective. Customers want to see the associated improvements in these processes for sure!

- Any project tasks that are incomplete at the end of a Six Sigma project turn into dissatisfaction of the customer. Such dissatisfaction is due to the mismatch between the customer CTQ requirements and the project results. These satisfaction points can also be caused due to the miscommunication between the customer and the Six Sigma team.

Example 4.6 A Kano model can be designed with measurable data analysis results after collecting data from users of a certain network to determine what customers expect more from the network. Possible improvement areas can be determined in the network processes and the degree of customer satisfaction can be identified (Fig. 4.15).

Chapter 4 Exercises

Exercise 4.1 Find an example (or sketch one) of a VSM that outlines a cybersecurity operation. Explain the specifications and the operations on the VSM.

Exercise 4.2 Find an example (or sketch one) of a spaghetti diagram that outlines a cybersecurity operation. Explain the specifications and the operations on the VSM.

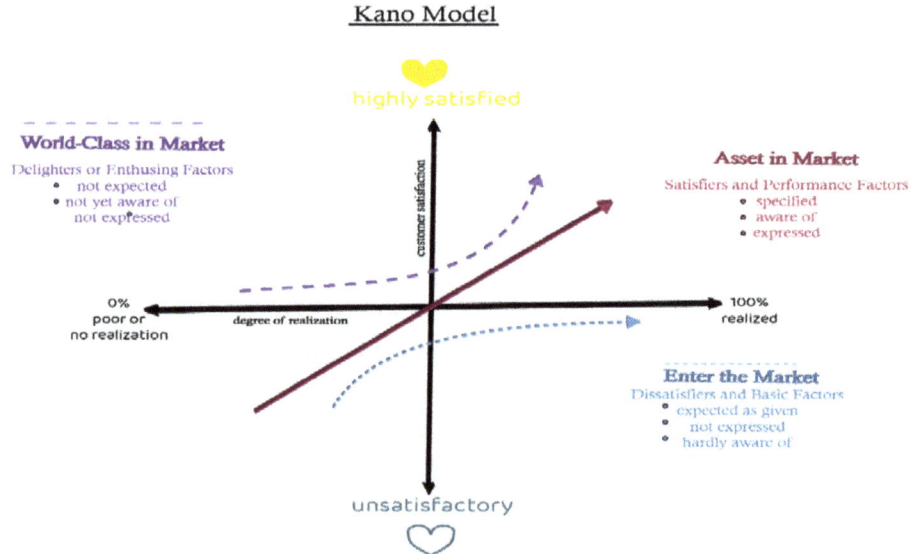

Fig. 4.15 An example of a Kano model

Exercise 4.3 Find a recent cybersecurity incident that had a major impact on an organization. Define the Six Sigma project associated with this incident. What are the key IPO steps?

Exercise 4.4 Define a cybersecurity real-life problem that is typically seen in real-life in the way that DMAIC approach requires to define the problem.

Exercise 4.5 Design a functional flow diagram for handling a specific cyber-thread that an organization may face. What are the business units involved in this handling process and what can be improved?

Exercise 4.6 Find a recent ransomware real-life event and explain the SIPOC component steps of the incident.

Exercise 4.7 Sketch a business flow diagram of a recent cybersecurity incident. Briefly explain the VA, BVA, and NVA activities with the business units involved in this thread handling process.

Biobliography

1. Framework for Improving Critical Infrastructure Cybersecurity. https://www.nist.gov/cyberframework
2. Tokgöz, E. (2024). Quality & lean six sigma applications for industrial engineers. Springer Nature, Switzerland. https://link.springer.com/book/9783031557392
3. Tokgöz, E. (2024). Quality and lean six sigma for engineering technicians. In *Synthesis Lectures on Engineering, Science, and Technology*. Springer Cham 978-3-031-44033-5. https://link.springer.com/book/9783031440328
4. Tokgöz, E. (2025). Artificial bee colony optimization techniques' utilization for intrusion detection systems' analysis. In *4th IEEE International Conference on AI in Cybersecurity (ICAIC) proceedings*. https://ieeexplore.ieee.org/stamp/stamp.jsp?tp=&arnumber=10848880

Measuring Variables for the Defined Problem 5

> Live it, learn it, collect it, store it, test it, fix it, collect
> it, retest it, confirm it, store it, maintain it, finalize it,
> document it.

Measurements are key to the success of a project for data analysis on the measurements to be implemented correctly therefore the Measure phase of DMAIC that contains data exploration, collection, and assessment is critical to the success of the Six Sigma project. Measure phase needs to focus on the data only relevant to the scope of the project since otherwise the project would have wasted time itself. There may be data sets collected prior to the Six Sigma project, however the data collected may require further processing for the purpose of the Six Sigma project therefore cleaning and editing the data may be necessary. Understanding the detailed nature of the project's scope, the relevant business units and entities, and the dynamics of the existing system are extremely for making important data collection decisions as a part of the Measure phase of DMAIC. Given a set of security metrics that can be appropriate for applying as a part of the measurements, a cybersecurity program establishes a common vocabulary with all technology support processes within the organization. It simply links business processes and cybersecurity processes that allow us to view cybersecurity risks as business risks.

Every system has its own rules and as a cybersecurity professional you might know the small cybersecurity space that you are stepping into with your theoretical knowledge and yet you may also have some practical knowledge in the area that you are about to explore. The most important question in this section is the following:

> What is the infrastructure of the cybersecurity system that you are about to start working on and how can you understand its nature to be able to analyze and improve it?

If you have not collected the data, the next most important question is

> How can I collect the data to be able to analyze and improve this system?

or if there is a collected data set shared with you then

> How can I understand the nature of the collected data?

Start looking into the details of what is needed and how we can identify them. These are very important questions that drive the entire project's scope and outcomes. Sometimes you need to live in the world of the collected data (or the data to be collected) to learn and understand its dynamics. If you don't have the gut feeling of how data framework is structured, then you cannot necessarily move on to Analyze and Improve phases. Your experience and prior knowledge may take care of this part, however even existing data for the same system may have different nature. It is always important to approach data collection methodology with caution.

In this chapter, we will assume that the data has not been collected for you (or shared with you) as a cybersecurity professional and you need to collect and measure it by using the techniques that are specific to cybersecurity data framework as well as general six sigma concepts. Even if the data is collected for you or shared with you for the project, you can understand or "question" and measure the nature of the data by following the steps explained in this book. Questioning the nature of the data is very important when compared to the end goal of your project. It is likely that the end goal of the project and data collection stages appear to have a big gap between them therefore everything might appear vague; however, you will learn how to close this gap in this book.

Our first goal in the first section is to cover the Measure phase basics and statistics that are essential to know as a part of measurement planning. Even though we do not deal with analysis of the data collected in the measure phase, the general software and online resources that can be used for six sigma analysis need to be known and searched for (if not known) in the measure phase. This is because the data storage space needs to be identified to analyze it later in a meaningful way by using the appropriate software packages or web-based resources. For instance, the extreme amount of data stored in Microsoft Excel might cause Excel to crash during the analysis if you plan to use VBA as the analysis method. Then, it makes sense to collect the data in MS Excel and analyze it by using another software package. There are many more technical details to be covered later in the corresponding section *General Software & Online Resources for Six Sigma Analysis*. Noting that you know the basics on measure, statistics and general software and online resources, we move onto specific cybersecurity measurement techniques and their applications. These applications will be particularly helpful to you noting that they are achieved after years of experience. There could be different modalities of data collection depending on the nature of the project:

- **Existing Data.** The data might have been collected and stored in a database.
  - **Repetitive.** The data can be collected again because the system under consideration is currently working (or would work again for your project) and it needs to be improved.
  - **Non-repetitive.** The data collected from the system to be improved cannot be repeated because it is either too costly or the entities in the system are unavailable.
- **Collection-Needed Data.** The data might need to be collected from an existing resource or system in place and it is possible to recollect the data at any given time.

In some projects, if the data is shared with you, the nature of the data shared with you may have issues that would not help you to analyze it:

- There might be an extreme number of errors and empty spots, and this would require extensive amount of data cleaning
- The input and output considerations may not match the end goals of your project
- It may not have all the details in the way you want to collect your data.
- The data may neither have explanations of the data collection procedure nor have any information on the methods, variables, techniques, and units.

All these issues may force you to either collect data yourself or request the data to be collected again based on the measurements you have and guidelines that you provide because of surveying. It is extremely important to remember at this stage that the nature of the data drives the outcomes of the project. For instance, suppose the population has 100,000 data points and a person decided to collect 10 sample data points because it is easy to do so, it sounds like a reasonable number, and it would be cheap. Anyone with this mindset or data collection approach would have a problem with the outcomes of the project because the sample is only 0.01% of the entire population and the results would not be representative of the population. You may then naturally ask "what is the right number of sample size and how can we collect it?" which is the question that we answer in the next section with basics on statistics we need to apply in the measure phase.

5.1 Measure Basics and Statistics

In this section we will cover the essential components of the Measure phase that are needed to move onto the Analyze phase. There are decisions that need to be made at the beginning of the Measure phase, and the first one is how to plan the project measurements. The highlighted components of the measurement are

- The key performance indicators that need to be measured,
- Understanding the critical process components to reflect them on the measurements,
- The needs for measurement system analysis,
- Displaying the needs of the customer(s), project, and cybersecurity in a meaningful visual manner by using Measurement Matrix,
- Statistics that we will need for measurement,
- Demonstrating measurements to make the best out of the collected data.

You always need to bear in mind that six sigma values the customer the most and the goal is to make the customer happy at the highest level. The bare minimum would be to fulfil the customer's project expectations.

5.1.1 Project Measurement Planning

The first and foremost important part of the Measure phase is planning. We start this chapter with project measurement planning to understand the nature of possible sampling techniques. What follows is the surveying and data collection components for the six-sigma project. Many six sigma professionals or cybersecurity professionals do not necessarily know how to survey and determine data collection components which are important parts of data collection methods in some projects. It is crucial to remember that communication matters, however efficient communication matters a lot more! We will look at ways to implement data collection planning.

5.1.1.1 Sampling Techniques

Gathering the correct data by using the right sampling techniques is an important step in the Measure phase of DMAIC. The data source and the gathered information from a population need to be a representation of the Six Sigma project's goals. For instance, if we are considering thread detection analysis then the associated population should be considered first and then additional locations where the thread data may need to be collected needs to be decided next. The scope of a Six Sigma project typically aims to focus on the sample of the data rather than the entire population, therefore the right sample choice that represents both the population and helps to solve the Six Sigma's defined problem is essential. In addition, the sampling technique followed to collect the sample depends on the purpose of the project and characteristics of the population. In addition, the size of the sample matters for reliable and accurate data analysis results. The two major categories of data collection techniques are the following that also have several subcategories to be explained in this chapter.

- Sampling with random selections
- Sampling without random selections

5.1 Measure Basics and Statistics

Sampling with Random Selections

Sampling through random selections from a population is followed in this categorization by assuming that a certain number of data points need to be selected randomly by following a random approach without any rule applied to the random selections. We categorize Random sampling into three to be explained next.

(a) **Simple Sampling:** The simplest method of sampling is giving equal chance for each data point to be drawn from the population.

Example 5.1 Suppose the goal of a company is to find out the satisfaction levels of the customers from the use of a network within a security system that is an important factor in customer satisfaction. If there are 20 million users of the system in the world, one million people randomly chosen can be a sample from the population. Each person is equally likely to be chosen for the sample data.

(b) **Stratified Sampling:** If a population is subdivided into groups, then random choices of data points from each subgroup within the population are conducted in this sampling strategy.

Example 5.2 Considering Example 5.1, if the customer data needs to be collected from 3 different continents of the world, then random choices from these three continents based on their user populations can be used. Suppose the following are the corresponding user populations.

- Continent 1—3 million users
- Continent 2—2 million users
- Continent 3—1 million users

Therefore, the user population in these three continents is 6 million. If the company wants to find out the satisfaction levels of 5% of the customers by stratified sampling, then the following are the sample sizes and the corresponding continents:

- Continent 1—150,000 users
- Continent 2—100,000 users
- Continent 3—50,000 users

Hence, Stratified sampling technique employs Simple sampling is used while collecting data from each continent.

(c) **Cluster Sampling:** In the case when a population may have (or needs to be split into) groups we can apply Cluster sampling as the random sampling technique. Hence, a population should have groups for sampling with the selection of groups implemented randomly to sample within the groups. The grouping criteria should be reasonable if random grouping is applied to form the groups within the population.

Example 5.3 Returning to Example 5.1, the company can split 20 million users into random groups of one million that form 20 subgroups and select one of the subgroups to finalize the selection of the million users.

Sampling without random selections
As the name suggests, sampling without random selections does not require randomness during sampling. This approach is particularly useful when the method of drawing from the data is specified by the project's scope.

(a) **Sampling with Specific Topic:** This sampling technique focuses on identification of the characteristics of a certain topic in the population and choosing only members of this group's population for sampling purposes.

Example 5.4 There are many cyber-threads and if the Six Sigma project focuses on a certain threat such as phishing attacks, then the sampling is designed accordingly. It is essential to keep in mind that we may still need information about the other threats as we may need to have comparative results with other threats.

(b) **Sampling on Subpopulation:** This sampling techniques focuses on determining a subpopulation first to be able to sample from the subpopulation; This can be accomplished by choosing a specific number of members of the population in equally divided time intervals.

Example 5.5 Suppose our goal is to measure and improve the network to be able to improve the network package flow within a network and we want to collect data every hour. There are key metrics to focus on such as bandwidth, latency, packet loss, and other network-based issues that cause bottlenecks within the network system. There can be solutions produced such as optimizing network topology, upgrading hardware, managing network traffic, and network traffic congestion reduction are some to be considered. The entities to be focused on for measurements need to focus on the highest traffic occurring locations therefore the sampling would take place in the associated subpopulation selected. Noting that the subpopulation size would be too high, sampling is necessary from the most congested network areas during specific days and times.

(c) **Systematic Sampling:** Individual samples are collected by following a specific pattern during this sampling technique.

5.1 Measure Basics and Statistics

Example 5.6 Suppose a certain vulnerability is detected and it follows a certain pattern within a system. Collecting sample data that relates to vulnerability systematically during a certain period would help to have data sampled systematically to help identify certain patterns of the data during this period over a larger period.

5.1.1.2 Surveying and Determining Data Collection Components

Data collection though surveying is one of the best ways to collect data for Six Sigma project. The typical use of surveying in real-life applications may indicate the use of typical or ordinary surveys that are collected by agencies; however, we are referring to a project specific survey that is designed by the Six Sigma team to be able to collect project related data. One of the best strategies of surveying is to prepare survey questions that are easy to understand with short statements and option selection if possible. Open-ended responses are also needed in such surveys to collect detailed information from users who would like to provide more data. In addition to such surveying, your communication skills matter in which case you may need to communicate effectively with anyone who could provide effective feedback. During such efforts, having a good understanding of how the data would look like after completing the data collection is needed because otherwise you could end up having a data set that is hard to analyze or hard to attain meaningful outcomes that relate to your Six Sigma project. This requires a comprehensive understanding of how systematic mechanisms work through your observations. In some Six Sigma projects the data set may be given to you as a spreadsheet, and you may have to use the given data set during the Measure phase with the focus on only the data set. Such a pre-collected data set may or may not contain the information that you need for your Six Sigma project therefore you may need to have either additional observations in the system because such a data set cannot be recollected, or you may need to re-collect the data if you can if it is not too costly. If it is not too costly and if you can recollect the data, then it is best to recollect data through sampling. If it is too costly to recollect the data and if the data shared for the Six Sigma project is not going to help for solving the defined problem for the Six Sigma project, the stakeholder(s)/customer(s) should know this fact.

To start preparing a survey, you need to follow a top-down approach for breaking tasks down to their components with the focus on the Six Sigma project. As an example, let's consider the data collection from three different server groups, and all server groups have their own characteristics that need to be incorporated in your survey. The network administrators of the systems of the company may need to be asked questions aside from the server data. Your survey questions need to be structured according to the differences in groups and their characteristics that need to be observed with the Define phase goals of the Six Sigma project: Brainstorming and consulting others to have their opinions of outside experts and Six Sigma team members can be highly value adding to the survey and the project results overall.

Initiating your data collection with a top-down approach that incorporates a data collection plan would be highly valuable. Keeping in mind the end goal of the survey and the associated data to be collected, the survey collection strategy can focus on separating the groups of people into subgroups with each subgroup receiving their own questions that relate to their own duties. For instance, the focus of survey questions can be on the following to understand how a system functions.

- Facts that relate to the system
- Each group's expectations based on the expected standards of the system,
- Performance considerations (such as overperformance, underperformance etc.)
- Expectations of the personnel from the system
- Issues that are experienced in the system
- Concerns related to the system

One should always remember that the persons that operate systems can be the ones who would be able to tell you issues, challenges, concerns, systematic behavior, and how the system can be improved therefore communicating with them and getting their opinions should be one of the highest priorities of the Six Sigma team. Such data might be the best data you can collect however it is also important to remember that an operator might be biased due to personal reasons therefore the data would still need to be collected from the system and the associated analysis should be confirmed. Matching analysis of the measured data from the system and the system operator would be a confirmation of the results. You can also collect information on system operators' best practices and the ways that they handle data and arising system-based issues. This is because those who work on the processes (such as operators or IT) may now how to solve the problem you are investigating however somehow they may not be involved in the solution to the problem (e.g. These issues could be not having time to resolve the issue because it is a big project or the communication between the upper management and personnel may not be strong to find a solution to the defined problem you are working on.) Fig. 5.1 illustrates the grouping of employees with the desired survey needs, expectations, issues, and concerns that need to be outlined by the Six Sigma team to be able to share the surveys with the associated groups.

Some of the rules that need to be kept in mind include the following:

- The survey design needs to be simple enough so that the participants would easily complete it; Surveys that contain too many questions and open-ended responses can be challenging to complete.
- Short questions that respect a scaling system such as Likert scale of 1–10 (or 1–5) as well as yes/no type of questions with open ended responses would be beneficial.
- The questions should be stated clearly with appropriate wording with no follow-up questions from the survey-takers.

Fig. 5.1 Desired outcomes need to be determined if certain groups exist with the determination of survey needs, expectations, issues, and concerns

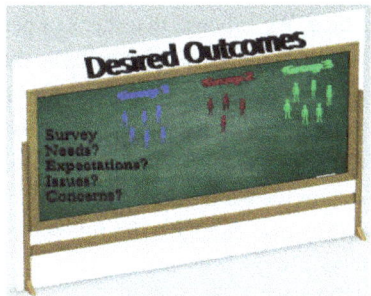

- The survey(s) need(s) to be administered initially to a test/focus group (which can be a randomly selected subgroup of each group) and a selected number of the subgroup members should be interviewed regarding the content, effectiveness, and length of the survey. The survey should be edited and finalized by using the feedback from the focus group members.
- The quality of the overall data collected also depends on the amount of data collected. This quantity depends on the project goals, regulations, periodic nature of the system observed, the statistical measurement considerations, and some other technical area aspects of the project.

Following appropriate steps with periodic project phases relate to the data collection planning methodology to be covered next.

5.1.1.3 Planning the Data Collection

Organizational skills and asking the right questions at the right time and right place are needed to plan data collection. An appropriate design of the data collection strategy is needed, and the output of this design methodology would be used for your Analyze and Improve stages later during the project. Improving this technique over time is a good strategy if the same (or similar) data with the same (or similar) goals of the Six Sigma project would be expected to be implemented later in the future.

Divide-and-conquer is one of the strategies that can be followed to split a process into its subprocesses for data collection. This can help to reduce waste in each subprocess and therefore reduce the waste in the overall process. As covered previously, administering a survey with the right questions relating to the end goal of a project to all relevant personnel is necessary and a divide-and-conquer approach can be helpful in accomplishing such a task. This is due to the need for understanding systems' capabilities and the associated measurement considerations. Splitting the data collection into several subphases based on the intensity of the data size (i.e., volume) and then splitting the process into subprocesses can help with the data collection as well.

For instance, suppose you want to improve the time you spend going to work from the moment that you wake up in the morning to the moment that you step into your office. We suppose that this is a periodic task that repeats every weekday during the same hours. You have similar patterns that you follow to be prepared for the work and there are certain tasks that you need to fulfill before leaving your house. Considering this as a Six Sigma project, the primary step is to break the steps down into subprocesses to identify the improvement opportunities based on your preferences. You would need to write each step of the subprocesses by breaking them down and derive logical outcomes. The process in this case can have the following subprocesses as potential options:

- Getting up from bed.
- Using the bathroom for preparing to the work—Further breakdown of this subprocess includes using the bathroom, showering, and shave/put make up on.
- Wearing work clothes.
- Going back to the bathroom to check how you look for any potential modifications.
- Checking if you have everything you need to leave the house such as car keys, house keys, computer, work security badge, etc.
- Walking to your car.
- Choosing a route to travel to work (assuming that there are multiple possible routes that can be followed.)
- Parking at work and walking into your office.

We can ask 5 W and How questions for all the above-listed subprocesses. For instance, let's take the step of checking if you have everything you need to leave the house subprocess. The associated questions would be the following:

- What items do you need to check to ensure that you are ready to leave the house before going to work?
- How do you ensure that you didn't leave any of the items before you left the house? (Are you keeping them in mind to collect them in a certain place because you can waste a lot of time walking around to find them.)
- Where do you place the items?
- Who may need these items in the house in addition to you?
- Why do you need all the items you take with you; Are there any redundant ones not needed?
- When is the right time to have all the items with you?

How and where questions in this list may relate the most as a part of time wasted and possibly reduce the time of your preparation for going to work. Other subprocesses can have their own 5 W and How questions that can contribute to organize your preparation. If you are following variational steps due to several reasons that you may need to collect your data to be able to analyze the best strategy and whether it would make sense to have the typical and repetitive tasks every day. Clearly this is your personal life therefore

5.1 Measure Basics and Statistics

it may not be desired to be this much lean when it comes to traveling to work due to psychological reasons! The proposed solution may look too automated with this example of a real-life situation which is more of a social event, however the real-life professional projects require such organizational methods to be followed to reduce waste.

Revisiting Example 5.5, we consider measuring and improving a network to be able to improve the network package flow within the network by collecting data every hour. There are key metrics to focus on such as bandwidth, latency, packet loss, and other network-based issues that cause bottlenecks within the network system. There can be solutions produced such as optimizing network topology, upgrading hardware, managing network traffic, and network traffic congestion reduction are some to be considered. We can ask a basic set of 5W and How questions on the data collection:

- Who collected and administered the data?
- Where did the data collection occur?
- How was the data collected?
- What was the source of data collection?
- Why is the method used for the way the data is collected?
- When was the data collected?

One other important consideration is the discrete and continuous nature of the collected data. Time is the typical example of the continuous data type (or variable) and countable entities such as people, number of servers, number of computers are examples of discrete data type (or variable.) Most of the data is collected in its discrete nature; even sometimes the time itself. This is a result of recording continuous variables as discrete values such as 1.472 s, however it doesn't change the continuous nature of the time.

In the case of data collection planning strategy, if the data is not collected for you, then you would need to have a clear plan of how to collect the data. During the data collection period, there are several key factors that need to be paid attention to. For starting the data collection and having a proper structure followed for the data collection, the following are several factors that can be followed:

- **Data Type.** Identify the type of data that needs to be collected. Examples of data types that we would use in this work include the following:
  - **Qualitative data.** Verbal or non-numerical natured data that includes the following.
    Interviews
    Surveys
    Data collection through verbal communication
- **Quantitative data.** Numerical data.
  - **Physical component data.** Physical data that we can collect from the system. For instance, a network system with different types of computers, switches, printers, servers etc. that takes place in the network can be grouped into different hardware

components as the physical component categories. Physical component data collection can be time and space consuming as well as overwhelming to the Six Sigma team therefore it needs to be done cautiously; Grouping such data into different categories needs to be also conducted carefully. People that work with the system can collect the data for the Six Sigma team.
- **Digital data.** The collection of digital data can include software related data such as numerical log data. It is essential to collect the relevant data to the project's scope as otherwise the volume of the collected data can be challenging to analyze and interpret. It is ideal to collect optimal data points and/or sets to be able to demonstrate the issue that needs to be demonstrated.

The data collection process starts with attaining permissions for accessing and copying the data regardless of the data collection method. The selection of storage and sharing rights with specific groups also needs to be clarified. Unattained rights can be troublesome to the Six Sigma team members if permission protocols are not followed.

5.1.2 Key Performance Measurements (KPM)

Performance is subject to the user and the requirements that need to be fulfilled for a project. The key performances need to be observed at both input and output of the project itself, and they are the variables that play the most critical role in the determination of the project. There are two important types of key performance measures to be covered in this subsection: input and output. One needs to remember the cost impact of each variable before determining the key performances to be used. This is usually the driving factor for cybersecurity consultants who work on fulfillment of projects. Each variable that needs to be used or determined in the project requires data collection and analysis. The more variables we have in a project, the more complicated it gets and yet possibly results in more meaningful outcomes. There is always a tradeoff between cost, number of variables used (both input and output), and the strength of results attained.

5.1.2.1 Primary Key Performance Input Variable (P-KPIV)
The six-sigma project starts with an end goal in mind and the corresponding output(s) requirement(s) to be fulfilled based on the system input. For instance, if you are baking a cake which is the desired output, you will need to go to the shopping center and purchase only the relevant items. Your shopping list for the cake is the list of key performance input variables (KPIVs) that shape up based on what you will be using for the cake. If your goal is to minimize the cost, then you will need to consider ingredients which are your KPIVs, and they are variables because there are different brands with the corresponding costs. The solution to this problem would be simple since there are a limited number of items and it is easy to find the minimum cost for each item by searching for them. If your goal is to minimize the cost but also maximize the quality at the same time, then

5.1 Measure Basics and Statistics

the input choice becomes complicated. If you don't know how each brand may impact the quality of your cake (and you don't want to collect your own data because it will take a long time and testing of each item) then you may need to look up online to see the comments on these ingredients. It seems to be a lot of work for baking a cake, but it might be worth doing it to enjoy the cake. Once you find the right ingredients, it will be your desired primary KPIVs to prepare the cake and make it sustainable. You could even improve this list by adding more ingredients at a secondary level and it would generate your secondary category for KPIV to be explained in the next subsection. Every time you add an ingredient, you need to repeat the same steps mentioned above for the original KPIV list and this will cost you more time, money, and energy but the return can be delicious cookies in the way you desire! In your project, it is essential to make the list of primary KPIV however it is optional to have the secondary KPIV category depending on how emergent you need to complete your project and how fast you need to get the results.

If we generalize the KPIV then we will need to think of it from set theory perspective. Say there are $S_{in} = \{x_1, x_2..., x_{k-1}, x_k, x_{k+1} ..., x_n\}$ number of inputs that impact the entire system which corresponds to the number of items; this corresponds to the items in the supermarket that may or may not relate to the output of the cake (initially we may not know it if we are uncertain about the ingredients of the cake. We need to find a subset of this set that consists of the ones that have the direct impact on the project's outcome and relate to the key performance measurement that we plan to utilize. Once we decide the ingredients of the cake then the corresponding ingredients form a subset SS_{in} of the set S_{in} (i.e., SS_{in} is chosen from the elements of S_{in}). This choice is going to be made from the project itself to be implemented. The KPIV choice can change depending on the scope of your project and your desired outcomes and measurement method.

5.1.2.2 Secondary Key Performance Input Variable (S-KPIV)

There are input variables that have some impact on the six-sigma project output, however they may not have strong influence on the output. We categorize these variables into the secondary KPIV group. For instance, considering the cake baking example, the taste of the cake might be better if we use an additional ingredient such as lemon juice (a secondary KPIV) however it is not essential to use it because the taste of the cake (the desired output) may or may not change based on this input. In your project, as you are working with a team, the team members may have different perspectives of the primary and secondary KPIV; therefore, it is the best if the team members would write down the primary KPIV and secondary KPIV to be able to discuss both categorize and finalize it with the agreement of the group members. Sometimes the teams do not necessarily have the knowledge to be able to make such a decision, especially if you are working as a consultant who is working on a six-sigma project, then you will need to prepare a survey for the representatives to fill out and collect data from them. We recommend preparing a survey and collecting data from all levels of the parties responsible noting that communication among the parties responsible is one of the most common issues seen in the workplaces that cause waste. For instance, consider an organization with both cloud and physical server operations. If the six-sigma project team consists of only system's manager, cybersecurity professionals, and software developers then the project would not be successful because there is likely to be a strong need of data analyst and IT for such a project with a possible involvement of finance department depending on the needs of the project; the project team leader needs to be as inclusive as possible to include staff at all levels that relate to the project. In some cases, for large corporations, the representative of the group (e.g., manager of IT) organizes meetings to gather information from the group and takes this information to the six-sigma team as the representative of the group. This team needs to determine the primary and secondary KPIV to succeed in the project. When compared to the Kaizen event approach, the six-sigma team's improvement idea applied in one of the areas of the project becomes the Kaizen event. For instance, suppose there are 30 different locations with identical work done in these locations. Suppose there are some differences in the way threat handling works in these locations, however these differences do not have a major impact on the six-sigma project output. The six-sigma team consists of.

- An IT manager that represents all IT professionals in all 30 locations
- A manager that represents all of cybersecurity professionals in all 30 locations
- A manager that represents all software developers
- A member of the finance department
- The system manager

This would be a very strong group that would determine the primary and secondary KPIV list. Representatives of the groups need to gather their KPIVs from their respective groups prior to the six-sigma team meetings. Primary and secondary key performance output variables (KPOVs) need to be determined prior to determining the KPIVs; The next two sections are reserved for the two possible KPOVs: Primary KPOV (P-KPOV) and Secondary KPOV (S-KPOV) that we consider in this book.

5.1.2.3 Primary Key Performance Output Variable (P-KPOV)

A successful project is always assumed to fulfill the desired expectations of the stakeholders for the project. These desired outcomes can occur at different levels. The primary key performance output variables (KPOVs) are the primary or mandatory outputs/expectations from the six-sigma project desired to be achieved by the stake holders. For instance, considering the cake baking example, your desired output is a good quality cake; however, this is your primary desired output and there is only one P-KPOV. Suppose you ask your family members if they want anything with the cake and they want to have something to drink with it as well as chips. Now you have three primary KPOVs. In this example we added more stakeholders, however it could be only you who want to have something to drink and chips with the cake.

5.1 Measure Basics and Statistics

In general terms, we define the set of all possible outputs from a project to be $S_{out} = \{y_1, y_2..., y_k, y_{k+1}, ..., y_n\}$. Each element of the set S_{out} is a possible output that can be achieved as a result of implementing the project within the company. We will use subset SS_{out} of S_{out} that outlines the specific project's outputs. For instance, if you are producing a machine, the set SS_{out} can consist of improving all possible outcomes that relate to improving the quality, reducing the cost of production, reducing the cost of shipping, reducing the cost of parts used for the machine etc. among all improvement opportunities. If you are working on improvement of a cybersecurity project with the goal of security improvement in possible network connections, then you can need a subset of all possible connections established throughout the system and focus on each chosen connection as a desired output to be improved.

5.1.2.4 Secondary Key Performance Output Variable (S-KPOV)

Project expectations can vary depending on the desired outcomes for the project as we mentioned in the previous section. After the six-sigma team determines the set of primary KPOV SS_{out} within S_{out}, the six-sigma team can determine the secondary KPOV (S-KPOV) set SSS_{out} within S_{out} that does not overlap with SS_{out}.

Revisiting the cake baking example, suppose you do not know yet about the toppings that you desire to have as a part of the cake. It could be only a plain cake as well though. You decided to go to the supermarket and look at several options to decide what you would have. The primary output in this case is the plain cake with no ingredients. The secondary output is what you may want to add to the cake, however it is secondary because it is not important to have the additional ingredients since your current most desired output is plain cake. Revisiting the cybersecurity project example that we have in the previous section with the end goal of security improvement in possible network connections, your P-KPOV might be improvement of the connections with the highest level of communication and security issues. The secondary KPOV set elements can consist of the connections with the second highest level of communication and security issues. The improvement of S-KPOV can depend on the time and resources available for the six-sigma team. It could be a very easy task if the communications and security establishments are similar in some of these connections therefore it might be possible to fulfill the S-KPOV as a part of the project; however, it could be also a very difficult task if the communications and security establishments are completely different. A decision needs to be made on P- and S-KPOVs to determine P- and S-KPIVs.

5.1.3 Critical to Measurement (CTM)

Measurements' critical components arise from several areas. While the customer's happiness with the results of the project is the key factor for understanding you were successful in the way you completed your project, this success achieved at the end depends on several

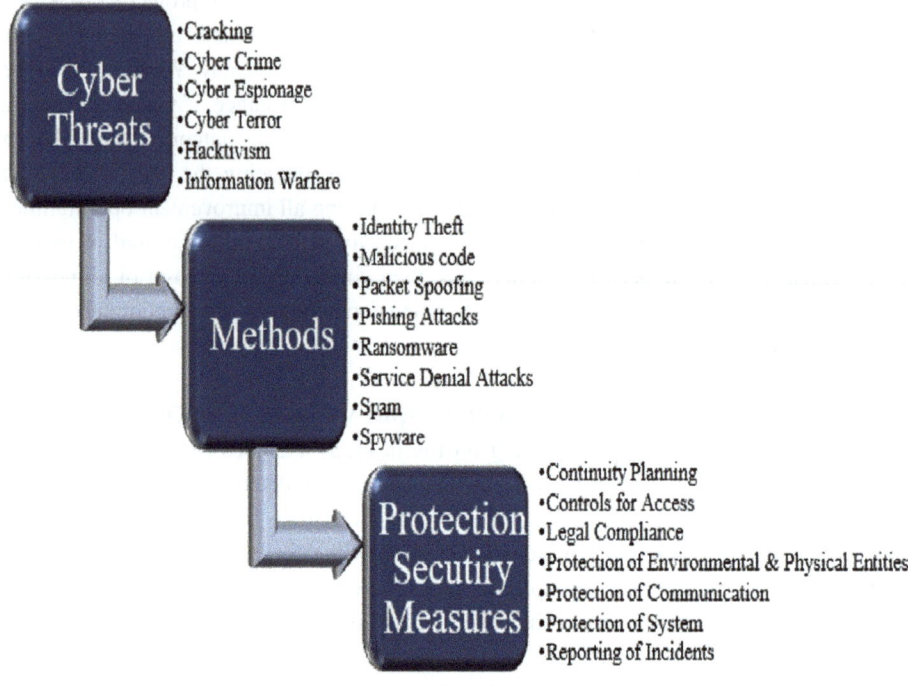

Fig. 5.2 Threats, methods, and protection security measurement considerations

other factors that need to be incorporated into your project otherwise you don't necessarily succeed even if your customer thinks you are successful. These critical components of measurement are process, quality, and overall satisfaction. Depending on the project's scope, overall satisfaction may require many components, however one of the satisfactions to be achieved should be the end-user happiness with the system in place. For instance, if the solution you attained is not desired to be used by so many workers in the workplace because it is cumbersome, even if it is theoretically excellent and loved by your customer (who can be the manager of the workers), then the system in place may not be used properly no matter how much you force it (Fig. 5.2).

5.1.3.1 Critical to Process (CTP)

Input-Process-Output (IPO) are the fundamental steps of a work to be completed in your six-sigma at high level and we covered the critical elements of the input (KPIV) and output (KPOV). The process that takes place in between input and output requires comprehensive coverage of the components that are critical to the process. The steps of the process for completing a work are critical for completing it in the right way. If we revisit the cake baking example, we covered the ways to determine the inputs and outputs and making decisions. Next step is to find the best method for completing all the tasks in between. You may think right away that we are baking the cake therefore the only step that takes place in between the inputs and outputs is the method to bake the cake however it is not. You will need to.

- look for ingredients missing in your desired input first,
- make a list of needed ingredients,
- determine the grocery store to purchase the items,
- determine the best efficient path to get to the grocery store,
- organize your list in a way that you will have no redundancies in the path you will travel in the grocery store without duplicating the paths that you are traveling through,
- complete your shopping
- return home by following the most effective route (shortest path, minimum traffic lights, etc.)
- follow the steps of cake baking (which could also vary, and you can find the best optimal way).

These steps require designing a flow chart and there can be sub-processes such as cake baking that would require their own process flow charts. What matters the most is determining the steps that are critical for your project goals and improvements. If you only care about baking the cake, then the rest of the steps are not critical to the process therefore you do not have to worry about them. The other processes that are not critical to our project goals can be a part of *Delighters* if they are improved noting that the project's end goal may not require accomplishing these tasks. Now let's consider the cybersecurity project with the end goal of security improvement in a network of connections that we discussed in the previous sections. The important process steps would be the critical steps where all connections are established throughout the system and security measures are applied. Each connection requires a process that links the KPIV(s) to the KPOV(s) to be improved. Outlining the steps of each process and identifying the process steps that yield to the waste would be the goal as a result of determining the critical to process components.

5.1.3.2 Critical to Quality (CTQ)

The quality conditions may need to be considered on any part of the project at any given step of IPO. Quality can be observed at many different levels in a six-sigma project depending on what aspect of quality you are dealing with. For instance, if we revisit the cake baking example, it is an easy example to understand the quality of the ingredients since they can have measurable outcomes established for quality. However, the ingredients that are critical for impacting the quality can be different based on the decision maker. For instance, you may use the tab water if you are comfortable with it however another person may choose to purchase water from supermarket. In some situations, it is a straightforward choice because of the quality driven by the product expectations. For instance, in aerospace applications, the expectations are straightforward based on the application and there are "musts" that need to be fulfilled with ISO requirements (such as ISO 9001). Even with these standards that set up the quality results, you don't necessarily meet the quality expectations of your customer. As a real-life example, there is an aerospace company that delivers items that fulfill ISO expectations; However, some of these items do not necessarily meet the customer requirements therefore these jet engine parts are returned to the manufacturer. The issues usually arise from "error term" perspective that we don't cover in this section. It is always important that the KPIV and KPOV are always driven by the choice of the customer along with the choice of variables that are directly related to CTQ conditions. Your customer may not be aware of the details therefore it can be your duty to include the KPIVs since it is now an important KPOV. In some cases, you need to design your own quality conditions. One of the best practices in cybersecurity (and in many fields of six-sigma applications) is to search for different measurement techniques established for measuring quality.

5.1.3.3 Critical to Satisfaction (CTS)

Satisfaction is a concept we touched upon previously; however, it needs more explanation. The ingredients that are critical for satisfaction primarily come from the customer's expectations; however, it is not the only condition to be satisfied. Customers may have certain expectations of quality, speed, cost, system design, reliability, performance etc., however there are other satisfaction areas you need to fulfill. We briefly talked about the quality standards and how these standards drive the critical quality decisions that are not necessarily decided by the customer. The customer may have higher expectations than the established quality requirements for the profession, however the quality standards still cost the customer because it costs the manufacturer. Revisiting the cake baking example, you are a customer to the grocery store that you shop, and they are trying to bring you a variety of brands with certain qualities. You personally would not know the quality of a brand unless you test it yourself or the brand fulfills certain standards for the product. The question is then the measurements needed for critical to satisfaction. Suppose you are going to use eggs for the cake and there are three options available for you that you would consider using for the cake: regular, cage free, and organic eggs. Even if you don't care about the kind of egg you use, the quality standards might be pointing out to use at least a cage free egg. It would be then up to you to choose cage-free or organic egg if you want to fulfill the requirements. You also need to keep in mind the stakeholders (i.e., your family) who will benefit from the product (or the output).

Revisiting the cybersecurity example related to the security improvement in a network of connections of a company, you may not have any standard protocol to follow in your six-sigma project and your customer may ask you to define critical standards that will yield to satisfaction. You are the one who then needs to determine the critical to satisfaction conditions for the system in place. How would you do it?

5.1 Measure Basics and Statistics

The answer to this question requires observations similar to cellular investigation of the human body but not as complicated! Here are critical considerations for this project's satisfaction:

- Review literature or search information on the practices that have been utilized in this area. If a method satisfies the others for a similar project, it could satisfy your customers as well (although different projects have different scopes, and your project may not have any similarities to other projects; your findings can give you an idea of the direction to follow.)
- Interview stakeholders about their interests to see what would satisfy them. The stakeholders in this case could be
  - the company owner(s),
  - the manager(s) related to the connected networks,
  - the network operators
  - the personnel and others accessing the network and utilizing the services.

There are critical criteria that would be important to be determined for satisfaction:

- Waste existing in the network and their role in satisfaction: This needs to be worked out on a detailed Value Stream Mapping (VSM) of the network in a simplified manner if it is too complicated. "Divide and Concur" type of approach can work in such a complicated system by which you can divide the network into several sub-networks and work out the details on the VSM; otherwise, you may overwhelm everyone taking in the project with a very complicated may of the entire network.
- Users existing in the network and their satisfaction: If the users of the connected network system are not going to be happy with your project's output, then the system would cause distressed users, and this is going to be unpleasant for the company. It might be your number one priority for the end goal of your project to keep in mind their interest and design an improvement strategy accordingly.
- Auditor satisfaction: Some of the systems' quality conditions require auditing therefore companies train their employees for the corresponding quality expectations and auditing. Your improvements need to align with quality standards and therefore the expectations of the auditors.
- Hardware satisfaction: The network system's hardware components would depreciate at a different rate over time depending on their usage in the system which is not under anyone's control. Preventive maintenance can is the maintenance/service on the system to prevent issues before they cause problems. It would be important to satisfy the right hardware conditions at any given time would be critical for satisfaction. Preventive maintenance would allow you to understand the depreciation value of failing components through data collection, analysis of the data and reacting on time before

the hardware causes any defect. An example of such an issue can be the outdated servers to run security systems for the associated hardware to run software to come over cyberthreat related issues.
- Software satisfaction: It will be critical to learn and understand the role of the software in the system when you are about to deal with a brand-new system. It would be critical to understand all steps in the network and determine whether a pre-existing software package can help with the improvement of the system or would the system require you to design a new software package for you (or an external company) to design this solution for the system. If such a brand-new software package would make a big impact on the network, then a cost–benefit analysis needs to be carried out to observe the trade-off between purchasing the software and its benefit to the system. This would again require data collection and measurement to have comparative results. If you plan to work with an external organization, then they probably have such data on the satisfaction however you still need to do your own cost benefit analysis. For instance, suppose you find a software during your six-sigma work that belongs to company X and the company sells it for hundred thousand dollars; it appears that the software may boost up your connected networks' speed by 20% and increase the cybersecurity risk by 30% based on the collected from a similar system that is tested by the company then it might be worth to invest in the software but the decision will be belonging to the stake holders of the six-sigma project.

In conclusion, it will be the six-sigma team members' responsibility to determine the conditions that are critical for determining critical to satisfaction conditions that depend on.

- the project,
- the waste existing in the project,
- the stakeholders' decisions,
- the additional requirements of the product or service.

There are many other examples that can be given to highlight the importance of different settings that would fulfill.

5.1.4 Measurement System Analysis Needs (MSA)

A detailed answer to the question "what is the needs to analyze a measurement system?" requires the coverage of Measurement System Analysis (MSA) as a part of Analyze phase of DMAIC therefore it needs to be observed as a part of the Measure phase first. The appropriate measurements and the corresponding results need to be observed in detail to be able to analyze the data in a correct fashion. The system to be used for measurement

5.1 Measure Basics and Statistics

may require additional considerations in place. There are MSA standards determined in several areas of interest such as aerospace that require measurement and analysis of inputs in a certain fashion. The traditional six-sigma MSA requires several stages of observations that need to be kept in mind:

- Linearity
- Stability
- Bias
- Repeatability
- Reproducibility

Figure 5.3 displays the summary on the relationship between these observation stages.

Linearity: Given an input in a system, suppose we sample measurements from the occurrences at a constant rate over time that we will call reference values. Linearity of the measurements occurs when the difference between the average of all the reference values and each reference value remains to be the same as a constant number (i.e. change in reference values don't change the difference.) It is ideal to collect as much data as possible to be able to have data to have a strong representation of the system. This sampled data collected can be compared with the standard (or expected) values for the input to the system. Such a standard can depend on one or more considerations:

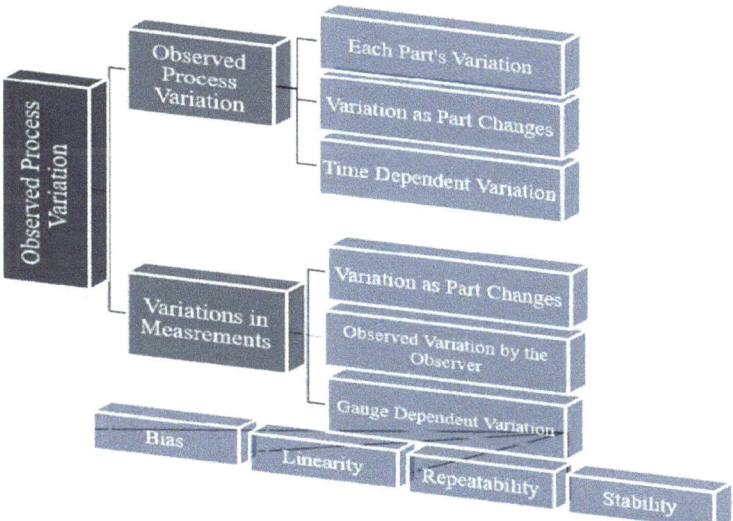

Fig. 5.3 Observed process' variation at different levels of security

- The expectations of the customer based on the expected output
- Expected industry standards such as ISO
- NIST-based guidelines
- Professional societies that have expected results and standards that are set up based on the existing data and experiences.
- Personal experience and judgement depending on the nature of the case, the data collected, and the expected outcomes outlined by the customer.

To standardize the collection of data, the sample data collection needs to follow a standard procedure, and the bias in the collected data needs to be minimized as much as possible by choosing and applying the appropriate sampling techniques properly. One of the most important aspects of numerical data is the need of the average value based on either the data collected points, and another one is the standard deviation that indicates how much the data is deviating from the standard (Fig. 5.4).

Stability: As the name itself indicates, stable means the system's data to behave in a stable manner over time. The expectation is to produce unchanging results over time for the system to act in a stable manner therefore any changes in the system can easily deviate the system from the expected behavior and values and might reach points where they may appear to be out of control. Such out-of-control points are indicators of irregularities in the system that can be due to vulnerability, a thread, or systematic malfunction depending on the system on hand. The data points for observing stability need to be measured over time and analysis of their behavior is critical for the system's success. Systems may have tolerance levels, and the stability of a system may depend on such tolerances that are typically known. Such values are called upper and lower control limits (that may or may not change over time depending on the system) and such values are used for determining

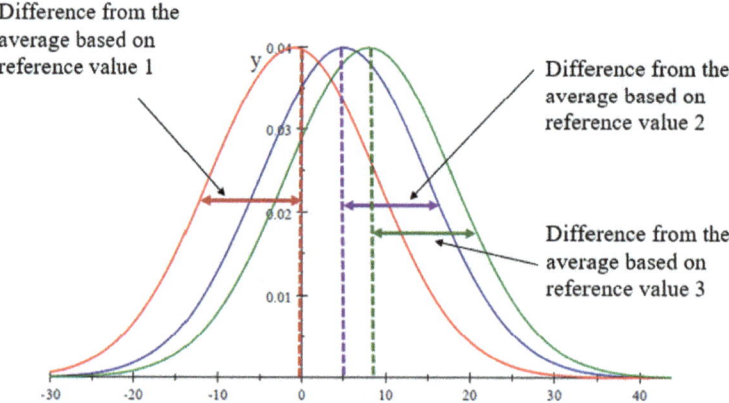

Fig. 5.4 The linearity of three reference values by using the average values for a normal distribution

5.1 Measure Basics and Statistics

significance of the changes on the stability of the analysis. For instance, a network may have 1000 users therefore the typical upper bound of usage is no more than 1000 which is the expected upper bound while the minimum number of usages may be 100 that can be considered as the lower bound. If such a network system has 2000 visitors suddenly then it may be the case that there is denial of service attack occurring. In the case of lower limit violation, the network may have issues with the critical functions as 100 IT network administrators may be the constantly logged in the system as users and the system may be down or one of such users may be experiencing issues. The upper and control limits are selected by the natural behavior in the measurement process and the average values may be the indicator of the typical behavior of stability. Figure 5.5 demonstrates the stability of a system's behavior during the first two times that the data is collected from the system. This behavior relies on the normally distributed data with stability relying on the centralized average value.

Bias. Inability to not follow the standards or expectations of a given system that relies on the expectations from the system causes bias. Therefore, the bias definition for MSA applications depends on the difference between the standard that is expected to be fulfilled by the system and the specific measurement. Bias values can be collected measuring the unusual behavior of the system by using the expected standard. For instance, a system may be experiencing three different threats which is the expected threat group. If another

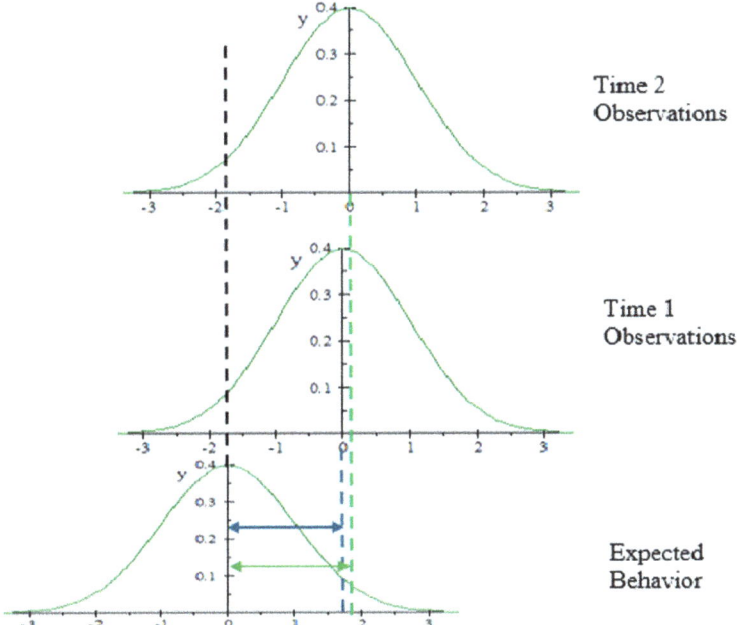

Fig. 5.5 Stability demonstration of a system based on two different times of data set collection

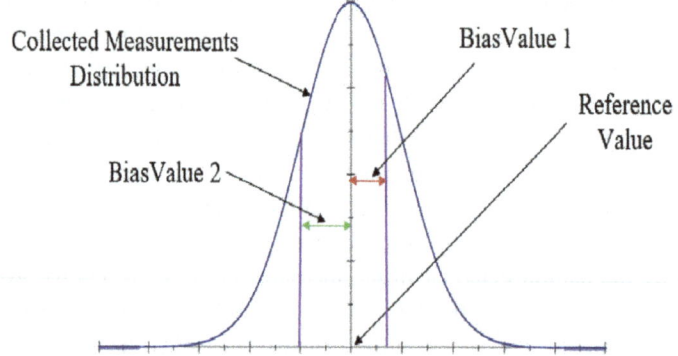

Fig. 5.6 A graph of bias in a system by using a data set that is normally distributed

threat is included in the system, suddenly then this increase causes bias in the system. Similarly, reduction by one threat in the system is also worth investigating with its causes and effects. The expected value can also be considered as the ideal case instead of the collected data, and in this example, the expected numbers of types of threats may be 2 therefore there are 2 more than expected threats in the system (Fig. 5.6).

Repeatability: Repeatability is the ability to produce the same results that relies on the variance measurements in a set of collected data based on the system that is observed under the same conditions. Repeatability of the analysis of the data is particularly very valuable in forensic investigations.

Systems that are typically expected to have the same behavior over time therefore the changes in these systems are indicators of unrepeating patterns. Such patterns are detected as unusual behavior of the system; however, such unusual behavior needs to be based on repeated actions on the same system. If the system changes, then its reaction to the same actions changes may impact the repeatability within the system. For instance, if you use a network system with different attributes every time you collect your data, you don't follow or repeat the procedure that you started with, therefore resulting in misleading collected data (Fig. 5.7).

Reproducibility: The ability to reproduce the results is an indicator of a system with unchanging characteristics. Given that the system is the same with unchanging characteristics, the reproducibility of the data collected by different observers in the system to check if similar or same conditions can be reproduced in the system is the key idea of this approach.

The Fig. 5.8 is an outline of the repeatability and reproducibility which is abbreviated and known as R&R in MSA applications.

5.1 Measure Basics and Statistics

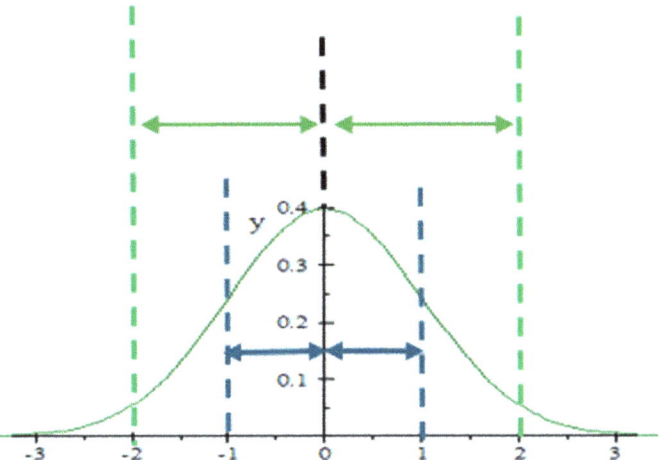

Fig. 5.7 The distribution of data representing repeatability based on the normal distribution

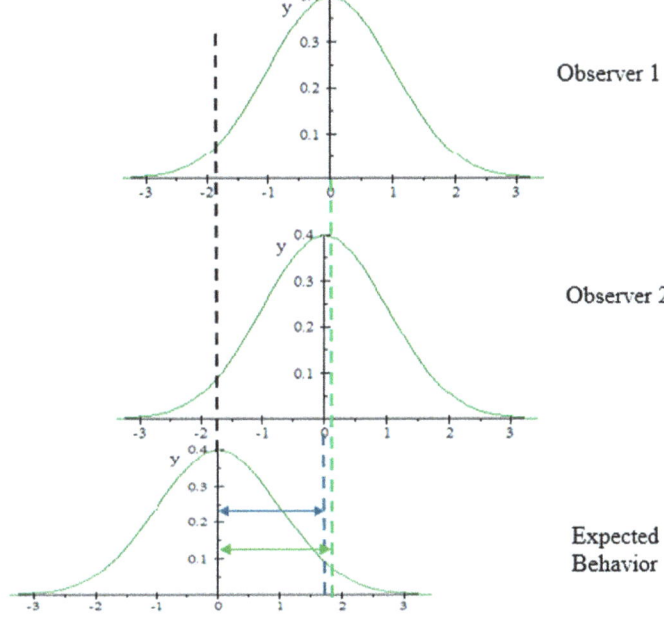

Fig. 5.8 A graph of reproducibility of the same (or similar) results by two observers

5.1.5 Measurement Matrix

Measurement matrix is an outline (or summary) of the relationship between input and output as a means of measurable outcomes typically represented in a table format. The inputs can be listed as rows in such a table with the outputs placed as columns. In this table, each row would represent an input, and each column would contain output. The quantification method used in this table depends on the weights to be assigned by the stakeholders, standard, Six Sigma team, and/or cybersecurity requirements. What follows is the explanation of such requirements that take place in the projects.

5.1.5.1 Stakeholder Requirements

The measurement matrix needs to be set up in such a way that it reflects the customer's expectations in a top-down approach. It would make sense (but not required) to list the customer requirements with highest to lowest importance from top to bottom. The customer requirements would rule the entire measurement matrix. If we say $N_R = \{r_1, r_2...r_k\}$ are the customer requirements, the rows of the matrix would consist of the elements of N_R with each requirement used one time. The P-KPIV and S-KPIV can be listed next to the corresponding row elements N_R and they can be listed once in each row but they can be listed in several rows. The outputs P-KPOV and S-KPOV are the corresponding columns in the measurement matrix, and they can be listed in the level of importance to the customer from left to right. The following figure is a 3D representation of such a measurement matrix. The weights need to be assigned to all KPIV and KPOV by the customer depending on the level of importance (Fig. 5.9).

Example 5.7 The following is an example of such a matrix that can be developed based on the data collected from the customer:

	Weight (Based on CTQ)			10	8	6	4	2		
	Output Requirements			Protection	Design	Value	Resilience	Strength		
	Actions		Inputs	Assigned Value to Actions					Total Value	% of Total Value
1	Cloud	Action 1	Source 1	60 (60*10)	80 (80*10)	90 (90*10)	25 (25*2)	100 (100*6)	2950	45.74%
		Action 2	Source 2	90 (90*6)	60 (60*10)	70 (70*6)	25 (25*8)	100 (100*2)	1960	30.39%
		Action 3	Source 3	10 (10*4)	80 (80*10)	90 (90*6)	10 (10*6)	50 (50*2)	1540	23.88%
2	Personnel	Action 1								
		Action 2								
3	Network	Action 1								
									6450	100%

5.1 Measure Basics and Statistics

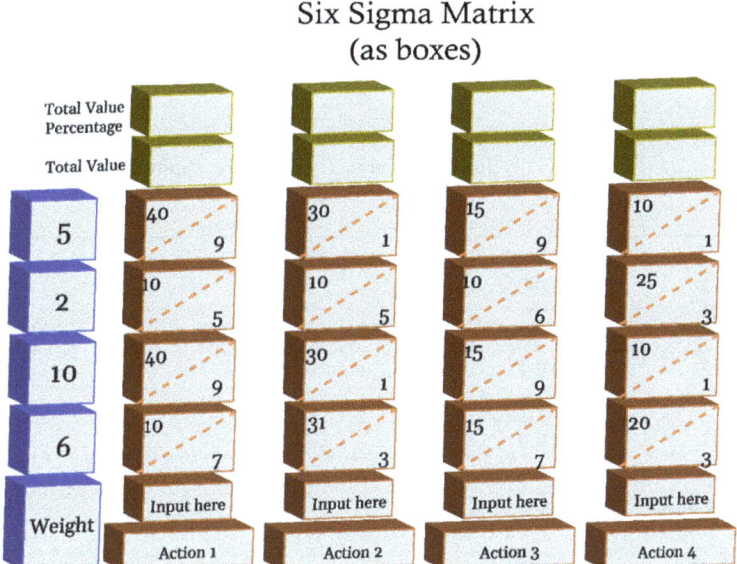

Fig. 5.9 Six Sigma matrix with values/weights assigned to CTQ conditions and operational actions

5.1.5.2 Business Requirements

The business requirements would cover all requirements that are essential for the project from a business perspective and not covered as a part of the customer requirements or the cybersecurity requirements (to be covered in the next section). There can be quality, finance, regulation, etc. related requirements that are carried out on the six-sigma project from the business side. The six-sigma team may have the corresponding professionals in the team, or the six-sigma team may need to get the essential information from the company representatives responsible. A measurement matrix like the one explained for customer requirements can be designed to analyze the business requirements.

5.1.5.3 Cybersecurity Requirements

In addition to the customer and business requirements, we categorize cybersecurity requirements as a separate category for designing a measurement matrix to make sure that it gets the special attention it deserves. The Cybersecurity measurement matrix would be expected to have the same outline of customer requirements' measurement matrix with the focus on cybersecurity rules outlined by the focus of the project. The structure of the cybersecurity measurement matrix follows the same guidelines of the customer requirements' matrix and same structure. Once the design of the three matrices is complete, the remaining task is to highlight the sum of the highest weights in each category and determine the measured results. The numbers would indicate the impact of the customer, business, and cybersecurity requirements in such a way that the decision makers would be able to decide which ones have the highest or critical importance to be focused on the three matrices designed. For instance, suppose we revisit the example of improving cybersecurity of connected networks. The customer requirement matrix has the following design:

		Design	Secure connection	Hardware	Software	Access	Total value & percentage
Customer Cybersecurity Improvement Requirements	Weight (10-point scale) →						
1. Network 2	P-KPIV 1						
	P-KPIV 4						
2. Network 5	P-KPIV 6						
	P-KPIV 2						
3. General Network Access	P-KPIV 1						
4. Data Breach	P-KPIV 4						
5. Network 3	P-KPIV 9						
	P-KPIV 8						

The table is designed for customers' cybersecurity improvement requirements. These requirements would be based on the cybersecurity improvements in the priority order of Network 2, Network 5, general network access, data breach and Network 3 among the 5 networks that exist in the system. Primary key performance 9 has higher importance for improving software of network 3 than P-KPIV 8. P-KPIV 4 stands for the strength of user password choice and it is common for Network 2 and data breach. There is no secondary level KPIV. The primary level KPOV in the level of importance are design, secure connection, hardware, software, and access.

5.1.5.4 Metric Requirements

KPI assessment and metric considerations to measure cybersecurity incidents go hand in hand. The possibility of measuring cybersecurity may be debatable however cybersecurity practitioners do use metrics for decision-making. It is essential to understand that cybersecurity metric assessment and the associated measurements change locally therefore prediction of incident occurrence within a certain location, or multiple locations can be predicted, and they can be identified up to a certain level of accuracy. Several attributes of the environment can be used to predict conditions of the environment, and the use of such attributes is generally successful when it comes to predicting large-scale issues that may occur. Therefore, in cybersecurity, even though the next attack location cannot be identified with certainty, it is possible to understand which systems (that store valuable

5.1 Measure Basics and Statistics

information for access) would fail on access control with successful attacks. The following are possible metrics that can be considered for cybersecurity measurements.

1. Actual versus attempted security incidents
2. Average detection time
3. Average response time
4. Average containment time
5. Identification of intruding devices on the network
6. Patching cadence and effectiveness
7. Effective employee training
8. Benchmark data that can be used for cyberthreat handling
9. Compliance with security audits
10. External risk and compliance

Widely used categories of security metrics are briefly described in the following table. An attack landing pattern and the associated metrics designed to identify such attack patterns are very much locally variable therefore their predictivity is just like a weather report that varies with varying locality of the location.

Categories of cybersecurity metrics		
Category	Metric description	Example
Performance	Metrics that demonstrate capability to accomplish system functionality	Transaction approval decision for audit trails of maker-checker authentication
Activity	Performed work activity	System access authentication
Vulnerability	Susceptibility to known threats	New vulnerabilities in the system
Adversary	Adversary based on motivation, justification, and resources	Cybersecurity index data
Remedy	A failed cybersecurity control objective's progressing status	Digital storage inventory scans for unencrypted personally identifiable information
Configuration	Digital assets used within an inventory used for secure configuration	Desktop Patch Status
Monitor	Tracking the security process steps for determining their fulfillment	Access to privileged credentials prior to privileged access usage
Periphery	Log data assessment based on the external organizational activities within the network boundaries of the organization	Failed network access attempts
Resilience	Metrics that demonstrate system ability to recover from harmful impact	Disaster recovery failover procedure timing

Six Sigma would require understanding the weaknesses that exist within the SIPOC expectations of the cybersecurity system and the associated wastes occurring within the waste categories outlined previously. For each waste category, appropriate KPIs need to be determined for collecting data and the associated KPI measurements with a specific unit to be used for the measurement. The data measured with the associated cybersecurity metric can be used for determining a correspondence to the waste category. Defining the measures and metrics required to properly analyze the waste categories is critical to identify and eliminate waste. Inefficiencies may be uncovered by activity and cybersecurity monitored metrics, in conjunction with performance analysis, however those metrics are not designed for six sigma application purposes directly. Therefore, in cybersecurity, it is common that inefficiencies are tolerated at the expense of cybersecurity control objective coverage that is organization dependent, therefore variable in nature.

Information Technology Infrastructure Library (ITIL) offers a generic technology control measure. That standard defines efficiency measures in relation to the productivity of the product or service, its speed, throughput, and resource utilization. Average cost per incident resolution is an example of such an efficiency measurement. It is important to separate this category from performance and resilience that may also measure attributes of speed and service to demonstrate that control and performance objectives are met. A metric that measures efficiency of a cybersecurity operation is selected and such a metric would typically measure two alternative processes to identify if the same control objective is achieved and compare benefits from an economic standpoint. The following is a summary of possible metric category with the associated possible waste that may occur in a cybersecurity system:

- **Activity.** Utility protection, production defects within the defense system
- **Adversary.** Talent utilization
- **Performance.** Local location cyberthreat occurrence, access performance
- **Efficiency.** Skill waste, lack of standards' fulfillment, staff related gaps, superfluity, speed of threat response performance
- **Configuration.** Internal and external computational weaknesses
- **Monitor.** External activities impacting network
- **Vulnerability.** Defects causing vulnerability within the network and security systems
- **Remediation.** Acquisition of defense effectiveness.
- **Resilience.** Storage of correct and reliable security data, delay on responses.

There are differences and overlaps between the cyber waste categories and standard six sigma waste categories. It may be possible to make a connection between the two categories, however cybersecurity has different dynamics than the typical manufacturing and healthcare applications therefore we introduced a somehow different categorization of wastes for six sigma applications in cybersecurity than the typical wastes used in six sigma.

5.2 Testing Hypothesis Statements

Hypothesis testing has many benefits in not only forecasting future incidents but also investigating the likelihood of events. Hypothesis testing is applied in many different real-life scenarios that depends on specific cases. As a part of this testing, we suspect a specific outcome and want to find out whether such a hypothesis holds true or not. Reliable data and application of the right steps to test such a hypothesis is essential for the success of the testing.

The following are the objectives of this section:

1. Understand hypothesis testing for cybersecurity decision-making.
2. Test hypotheses application based on the average value of a normal distribution by using a Z-test or a t-test approach.
3. Test hypotheses by using the Variance or Standard Deviation of a normal distribution

There are a few questions that can be asked to work on hypothesis testing:

- How can we decide to choose one of the options between two opposite choices for data evaluation for hypothesis testing?
- What statistical method can be used for making such hypothesis testing decisions?

Noting that hypothesis means competing claims, we can use hypothesis testing as the decision-making procedure for data set analysis for comparative purposes. Statistics is the main driver of fundamental methods used employed for hypothesis testing; For instance, Confidence interval estimation of parameters is a fundamental application of statistics.

A statistical hypothesis is a statement about the parameters of one or more populations. Our focus in this section will be on the control limits by using confidence intervals from a data driven perspective with the corresponding measurable calculations with the following coverage:

- Construction of confidence intervals by using data set based on the Mean value of a normal distribution, if the data has a normal distribution behavior.
- Construction of confidence intervals by using the variance and standard deviation of a data set that has a normal distribution nature.
- Construction of confidence intervals on the sample portion of a population.

Estimation is one of the primary elements to be observed for hypothesis testing purposes and we need to ask, "How good is an estimation?" To answer this question, we use an interval estimate for a population parameter that is called Confidence Interval (CI). Precision and the associated estimation information is conveyed by the length of the time interval used. Data collection over a short time interval is likely to provide a data set to reflect the system in its ordinary behavior. It is important to keep in mind that the collection of such data is costly, and its analysis can be challenging if it has high volume.

A tolerance interval (TI) is another important type of interval estimate that is an indicator of what the system may allow to tolerate based on the observations. One use of confidence interval is quality control analysis and there are many other applications that also help to measure stable behavior of a system.

As we covered previously, Central Limit Theorem is an indicator of approximately normally distributed data in the case when we use a sufficiently large data set. This allows us to use the mean and standard deviation values of the data to be able to construct CI. Hence, in the case when we have a sufficiently large data set that is normally distributed, we can use the confidence and tolerance intervals for normal distribution applications. In the case of a normally distributed data set, 95% of the distribution falls in the following interval (with the average μ and standard deviation σ that are not necessarily known)

$$(\mu - 1.96\sigma, \mu + 1.96\sigma)$$

There is a need to use a potential error in each point estimate to form a tolerance interval for the distribution therefore the interval changes to the following:

$$\left(\overline{X} - ks, \overline{X} + ks\right)$$

Such a tolerance interval approximation is used in many real-life applications that have a normal distribution nature. It is usually the main interest to start with a sample data set that represents a population and derives approximate results for the population after collecting large enough data. Hypothesizing a certain outcome by using such data and determining whether the outcome holds or not relies on the hypothesis testing. Typical statistical measurements used for comparison are the following:

- Mean value comparison of two or more groups.
- Variance or standard deviation comparisons between two or more groups.
- Analysis result comparisons between samples of different sizes and proportions of the population.

By hypothesizing, there are several different strategies that can be followed to test the hypothesis that will be explained in this section.

- 1- and 2-sample t-tests
- Paired t-test
- Analysis of Variance (ANOVA)
- Chi-square distribution
- Equal variance testing

5.2 Testing Hypothesis Statements

The right choice of the hypothesis testing method that depends on the scope of the problem is an important part of attaining successful results.

5.2.1 One-Sample t-test

The goal of the 1-sample t-test is to compare a given value (i.e., hypothesized mean value) with the mean value of a population. The assumption for implementing this test is to have data that obeys normal distribution. This comparison is based on the p-value in which case the p-value is compared with the benchmark value 0.05 (i.e., 95% accuracy). In the case when the p-value is greater than 0.05 we know that there is no statistically significant difference between the mean value of the population that is assumed to be derived from the sample population and the given value. In the case when the p-value is less than or equal to 0.05 we know that the two values compared with each other are significantly different from each other.

5.2.2 Two-Sample t-test

The 2-sample t-test has the same foundation as the 1-sample t-test with the exception that we compare two populations; it is assumed that the data of the two populations are normally distributed, and the comparison is based on the p-value. Exactly in the way it is hypothesized for the 1-sample t-test, in the case when the p-value is greater than 0.05 we know that there is no statistically significant difference between the mean values of the two populations. In the case when the p-value is less than or equal to 0.05 we know that the mean values of the two populations are significantly different from each other.

5.2.3 Paired t-test

The goal of the paired t-test is to compare the mean values of two samples that serve the same purpose. The two sample tests are expected to be related to each other in which we want to determine the significant difference between the two entities. The confidence intervals of a paired t-test are expected to be tighter compared to other tests' variation analysis. The paired t-test depends on the p-value analysis.

Example 5.8 Suppose the goal is to design a cybersecurity decision support model for building an information technology security system based on risk analysis and the ISO/IEC 27,001 cybersecurity framework [16]. The goal is to help strategic policymakers in designing cyber security decision-support recommendations to determine the best steps in designing information technology security systems. The idea behind the built model is to map the priority value of threat mitigation based on the relative threat score against the relative evaluation score of the implementation of ISO/IEC 27,001 compliance. The key in determining priority recommendations for building an information technology security system based on the ISO/IEC 27,001 framework is driven by the mitigation priority value. In this case, the information technology security system recommendations can be tested by carrying out security attacks directly on the system being built. For such a case, the effectiveness of the methodology can be tested by using hypothesis testing to attain numerical measurable outcomes. To do so, fake cybersecurity attacks directly targeting the company's systems occur during the testing and applications to ensure the implementation of the selected security recommendations can overcome and mitigate security attacks and threats. The numerical evaluation using quantitative analysis shape security recommendations that have been built. This is a process that requires continuous improvement due to the evaluation of the systems in the ever-changing cybersecurity systems designed, therefore application of six sigma is natural. The numerical evaluation aims to find out whether there are differences before and after the implementation of security recommendations. One viable way of implementing statistical analysis is by using R programming language with a value of 0.05 degree of freedom.

The hypothetical testing relied on data collection conducted by distributing questionnaires. These questionnaires are given a weighted value to respondents who have a relationship with company assets related to the success of security recommendations against threats and security attacks that exist in the company based on test data from the results of the implementation of the recommendations. Validation of the questionnaire data (such as expert judgement method) before distributing to the respondents helps with the accuracy of the analysis. What follows next is building hypotheses based on the research problem based on the following KPIs:

1. Success for detecting and blocking *brute force*
2. Success for detecting and blocking email *Phishing using a link*
3. Success for detecting and blocking *Malware and Virus use for email Phishing*
4. Success in detecting *Modification File Configuration*
5. Success of Failover of *Power Supply Redundancy*
6. Success of Failover of *Network Adapter VSAN Redundancy*
7. Success in finding *vulnerability in the system*
8. Success to finding misconfiguration based on *Security Configuration—Compliance*

The following are the respondent criteria used during the data collection. The data is collected based on the person's function…

1. Directly responsible for the asset object
2. On the use of asset objects

3. For the Operational continuity of the asset object
4. Related to asset management of corporate regulatory compliance

The following is a hypothesis with the analysis conducted by using a paired t-test analysis:

- *Null hypothesis (H_0).* There is no significant correlation between threats to information technology security systems that implement and do not implement information technology security system recommendations on the value of the ISO/IEC 27,001 compliance evaluation index.
- *Alternative hypothesis (H_a).* There is a significant influence between threats to information technology security systems that implement and do not implement information technology security system recommendations on the value of the ISO/IEC 27,001 compliance evaluation index.

Data is collected and evaluated based on the conditions that existed before and after the implementation of information technology security system recommendations to evaluate the implementation of the ISO/IEC 27,001 framework in the organization. Paired t-test method is used for determining the significance of level differences between the two conditions of information technology security systems in organizations. In this case, the data population must be normally distributed and have the same homogeneity. Noting that both conditions are met, upon numerical application of the test (that we don't represent here and leave it to the reader to investigate), the Paired t-test on the data population can be carried out. The results of the significance test with the paired t-test method are shown to have a p-value of $0.002138 < 0.05$ therefore the null-hypothesis is rejected meaning that there is a significant influence between threats to information technology security systems that implement and do not implement recommendations for information technology security systems on the value of the ISO/IEC 27,001 compliance evaluation index.

5.2.4 Chi-Square (χ^2) Statistical Method

This method utilizes the following formula for analyzing the difference between the observed and expected values. It aims to observe if the difference between the observed and expected values are by chance for the KPIs (or variables).

$$\chi 2 \text{ formula}: \text{Sum}\left(\frac{\textit{Observed Value} - \textit{Expected Value}}{\textit{Expected Value}}\right)$$

Following the information and data shared in Example 5.8, we will see that Chi-square method is not applicable for the hypothesized case while the second example would indicate the method's applicability on the data for the associated hypothesized case.

Example 5.9 *(Chi-square & Fisher tests)* Given the same scenario in Example 5.8, the data for the following hypothesized testing is obtained from respondents' answers to a questionnaire consisting of 12 examples of threats that exist in the organization's information technology security system. The answer to the questionnaire has an objective to assess the criticality level of a security system threat to the organization's assets and business against the mitigation steps taken. The data population consisted of the threat criticality level before and after the implementations.

- ***Null hypothesis (H_0).*** *There is no relationship between threats to information technology security systems that apply recommendations and do not apply recommendations to the level of criticality of information technology security threats.*
- ***Alternative hypothesis (H_a).*** *There is a relationship between threats to information technology security systems that apply recommendations and do not apply recommendations to the level of criticality of information technology security threats.*

The following figure demonstrates the difference between the two populations' data evaluation. There is a clear decrease in the criticality level of threats to conditions before and after the implementation of security system recommendations (Fig. 5.10).

As can be seen from the graph above, critical threats with extreme and major levels were determined before the security recommendations implementation, these issues were not determined after security system recommendations' adaptation. A non-parametric statistical approach of Pearson's chi-square t-test method is used for testing and this approach is chosen due to the requirements for the parametric statistical approach not meeting to get significant value for the changes in the conditions of pre- and post-testing. This method resulted in a p-value of $0.0006605 < 0.05$ meaning that the null hypothesis was rejected.

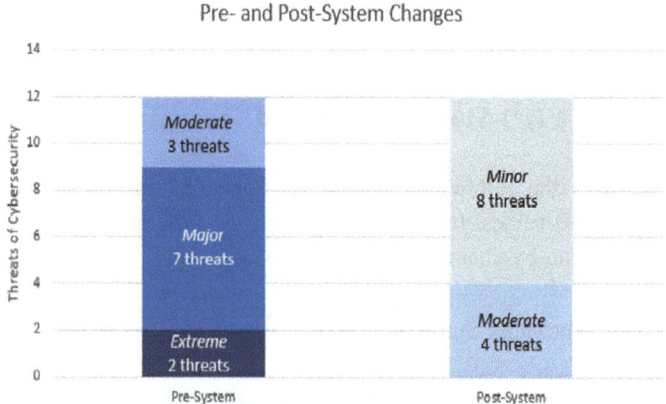

Fig. 5.10 Demonstration and pre- and post-systematic changes

5.2 Testing Hypothesis Statements

The p-value on Pearson's chi-square test may be wrong due to the poor accuracy that is known to be the shortcoming of the non-parametric statistical approach so that validation of the accuracy of significant values is needed by using another testing method. In such a case, Fisher test with the following equation can be used to overcome the shortcoming of the non-parametric statistical approach so that validation of the accuracy of significant values can be fulfilled:

$$p - value = \frac{(x_1 + x_2)! * (x_1 + x_3)! * (x_2 + x_4)! * (x_3 + x_4)!}{x_1! * x_2! * x_3! * x_4! * (Total frequency)}$$

where x_i values are the values in the contingency table.

Example 5.10 (*Chi-square test*) The following is the third hypothesis conducted on the information used for Example 5.8:

- ***Null hypothesis (H_0).*** There is no association relationship between systems that implement and do not implement recommendations based on ISO/IEC 27,001 for cybersecurity attack mitigation.
- ***Alternative hypothesis (H_a).*** There is an association relationship between systems that implement and do not implement recommendations based on ISO/IEC 27,001 for cybersecurity attack mitigation.

Responses of all 19 participants to the 12 questionnaires are used for the evaluation of the third hypothesis' data. The assessment's objective is to identify the effectiveness of the mitigation measures based on the ISO/IEC 27,001 framework after the implementation of information technology security system recommendations against cyber security attacks. The responses by all 19 participants showed an increase in threat mitigation scores from pre- to post-implementation based on the fake cybersecurity attacks. Using Pearson's chi-squared method, a p-value of 000,005,221 < 0.05 is attained meaning the null hypothesis needs to be rejected. The p-Value value in Pearson's chi-squared test may be wrong, so validation is needed using the Fisher test method, the p-Value value is 0.00000005658 < 0.05, which means the null hypothesis is rejected. So based on the two results, p-Value < 0.05, it can be concluded that there is an association relationship between systems that implement and do not implement recommendations based on ISO/IEC 27,001 for cybersecurity attack mitigation.

5.2.5 Development of Confidence Interval

The development of CI is helpful for understanding how confident we are that the given data remains within a certain range from a specified standard for the system that the data is collected from. Therefore, if we have a data set with numbers, it is not easy to understand if the points are away from the average of the data or where the data points stand within the entire data set. As an example, if we have 1200 as one of the data set values then we won't necessarily understand or know whether this is an extreme point within the entire data set or not. The way to observe such factors is by standardizing the data set that becomes a new random variable that represents the data points of the data set. In statistics, this new variable is known as the z-score defined by the following:

$$\text{New data point} = \frac{\text{data point} - \text{sample's average}}{\text{sample's standard deviation value}}$$

The following mathematical calculations help to find the z-values corresponding to the data set points mathematically based on the following assumptions:

- We assume that $X_1, X_2..., X_n$ is a random sample that has a normal distribution structure with unknown mean value μ and known standard deviation σ.
- The sample mean $\overline{X}$ is based on the normal distribution of the data with mean value μ and variance σ^2/n as we have n values in the data set.
- Standardization of the input values of the data set (i.e. x values) by using the z-score formula:

$$Z = \frac{\overline{X} - \mu}{\frac{\sigma}{\sqrt{n}}}$$

This formula helps us to identify the confidence interval for a given data set based on the expected percentage of accuracy.

- Z becomes the new random variable replacing X values as it has a standardized normal distribution structure. This standardization allows us to take data sets with variational inputs and standardize them into one structure. The corresponding formula for the entire population is known as the following:

$$Z = \frac{\overline{X} - \mu}{\sigma}$$

As can be seen in this formula, the sample size n n is not factored into the formula. As it will be covered later, the choice of n in this formula is useful for determining the right sample size and changing the sample size for determining the right parameters for expected z-value.

5.2.6 Confidence Interval Formula

Using the sample mean $\overline{X}$ of a random sample of size n from a normally distributed population with a known variance value of σ^2, we can calculate the 100(1-α)% confidence interval based on the average value μ by using the following formula:

$$\overline{X} - \frac{\sigma}{\sqrt{n}} * Z_{\frac{\alpha}{2}} \leq \mu \leq \overline{X} + \frac{\sigma}{\sqrt{n}} * Z_{\frac{\alpha}{2}}$$

with the upper 50α % of the standard normal distribution is represented by $Z_{\frac{\alpha}{2}}$. The following figure illustrates the Z values for both upper and lower $\frac{\alpha}{2}$ portions of the areas based on the left of the dashed line on the left and the dashed line on the right side of the right dashed line. The entire area between the distribution curve and the input axis is 100%. This region for the z-values make up α% of the entire area based on the two regions.

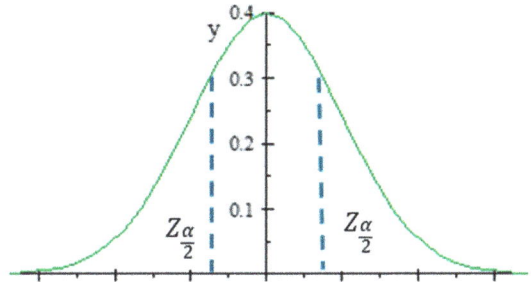

Example 5.11 Suppose 100 data points collected for the time it takes to respond to a threat that occurred in a system (in threats per seconds) and the data is collected in a data set that is identified to be approximately normally distributed. This normal distribution has a mean value of 10 s per threat, and a standard deviation of 2.5. It is an indicator of the response rate to the threat within the data set with the maximum percentage possible.

We can determine the 95% CI for the mean value (μ) of the threat as follows:

- n = 100
- σ = 2.5
- $\overline{X}$ = 10
- $\frac{\alpha}{2} = \frac{100-95}{2}\% = 0.025\%$ hence $Z_{\frac{\alpha}{2}} = 1.96$

Therefore, the 95% CI for the data set would be

$$10 - \frac{2.5}{\sqrt{100}} * 1.96 \leq \mu \leq 10 + \frac{2.5}{\sqrt{100}} 1.96$$

$$9.51 \leq \mu \leq 10.49$$

Interpretation. 95% of the threats in the data set are responded anywhere between 9.51 and 10.49 s.

5.2.7 Choice of Sample Size

The right sample size choice is very important for right and accurate analysis. If we are using $\overline{X}$ for the estimating of the average value μ, our confidence will be $100(1-\alpha)\%$ to ensure that our error value calculated of $E = \text{error} = |\overline{X} - \mu|$ does not exceed a specified amount error when the sample size is the following n:

$$n = \left(\frac{\sigma * Z_{\frac{\alpha}{2}}}{E}\right)^2$$

Example 5.12 *(Vulnerability Patching)* Returning to the vulnerability patching example introduced previously, and specifically "Production under 30" part, suppose our goal is to determine the number of data points needed for the 95% CI on μ for a mean value of 12 vulnerabilities and a standard deviation of 2. As an example, we choose the error estimation E as 20% of the length of the CI to determine n with $E = 0.2, \sigma = 2$, and $Z_{\frac{\alpha}{2}} = 1.96$. The required sample size is

$$n = \left(\frac{2 * 1.96}{0.2}\right)^2 = 384.16 \cong 385$$

Hence, 385 data points are expected under the specified conditions. We covered only a specific portion of the control limits of some one of the mathematical methods in this book; However, there are many other methods that control charts can be identified based on varying conditions.

5.2.8 Null Hypothesis

What is the meaning of mathematically hypothesizing, and how do we state and prove or disprove such a hypothesize? At the most fundamental level we need to first answer this question. For instance, it is possible to describe a server's ability to process and transfer data as a production capability by using probability distribution. Such information can be utilized to structure a hypothesis statement that can be tested mathematically. We let H_0 to be the **null hypothesis** that assumes the initial claim to be true while H_1 is the **alternative hypothesis** stating the contrary of the null hypothesis. For instance, let's say

5.2 Testing Hypothesis Statements

we estimate that this production rate is 600 emails sent per minute. Hence, we can start outlining a two-sided mathematical hypothesis by stating the following first:

H_0: $\mu=600$ emails per minute
H_1: $\mu \neq 600$ emails per minute $\rightarrow$ $\mu<600$ and $\mu>600$

One-sided hypothesis statements can also be produced as the following:

H_0: $\mu=600$ and H_1: $\mu<600$
or
H_0: $\mu=600$ and H_1: $\mu>600$

Hypothesis is a statement that needs to be proven or disproven therefore it needs to be a clear statement about the population or the distribution of the data. Such a statement is typically made in a meaningful way due to experience or experiments/tests during a Six Sigma project. Hypothesis testing allows us to test the significance of a statement that we suspect to be true, and it either proves or disproves based on the evaluation of data used. The statement can be significantly true or significantly false. Hence, incorporating hypothesis into the decision-making process is a procedure leading to a decision about the null hypothesis which is called a **test of a hypothesis**. Hypothesis-testing procedures need to rely on reliable data collected as samples of the system's representing components of interest. Such data needs to be consistent to either prove or disprove the null hypothesis, since otherwise the hypothesis cannot be evaluated by using such data to accept or reject the null hypothesis; If the data's insight is consistent with the null hypothesis' statement, then the null hypothesis can be evaluated by using hypothesis testing to be explained in this section.

Given that testing null hypothesis for an equality may be too restrictive for measuring a system's effectiveness, we can utilize a region within our suspected behavior that is called **acceptance region.** Such a region contains our suspected null hypothesis value, and it can be based on a tolerance that the system has (or we have) for system's functioning. If the results are not within the acceptance region, then they would be in the **critical region**. The equality taking place between the acceptance and critical regions are called **critical values** of the hypothesis.

Example 5.13 (*Hypothesis Testing*) Suppose that we collected $n = 100$ data points from a server that distributes emails, and we suspect that 600 emails are sent by the server per minute. Such a sample is just a small estimation of the population. We will not be rejecting the Null Hypothesis H_0: $\mu = 600$ if the acceptance region $590 \leq x \leq 610$ holds. The critical region would be $x \leq 590$ and $610 \leq x$. The critical values are 590 and 610.

5.2.8.1 Errors of Type 1 Category

Rejection of the null hypothesis H_0 when it is true results in Type 1 error.

Type 1 Error Probability.

$$\alpha = \text{Prob(type 1 error)} = \text{Prob(rejecting } H_0 \text{ while it is true)}$$

Type I error probability is also known as the **significance level, α-error,** or **the size of the test**.

Example 5.14 Continuing with Example 5.13, the deviation of the email server from the standard for the 100 data points collected as 100 emails per minute ($\sigma = 100$) with the average of $\mu = 600$. Supposing that we have approximately normally distributed information (which is necessary for the following calculations), we have.

$$\frac{600}{\sqrt{100}} = \frac{600}{10} = 60$$

Therefore, the probability of making a Type 1 error, based on the critical region considered in Example 5.13 can be calculated as follows:

$$\alpha = P(x < 590 \text{ when } \mu = 600) + P(x > 610 \text{ when } \mu = 600)$$
$$= 0.1587 + 0.1587 = 0.3174 = 31.74\%$$

Based on our acceptance region, there is 31.74% chance of making a Type 1 error that leads to the rejection of the hypothesis H_0: $\mu = 600$ emails per minute to be sent by the server.

Using z-values for Type 1 Error Calculations: The use of z-values for calculating probabilities makes it easier to achieve results. By using the critical values 590 and 610 we can calculate their corresponding z-values first as follows:

$$z_1 = \frac{x_1 - \mu}{\frac{\sigma}{\sqrt{n}}} = \frac{590 - 600}{\frac{100}{\sqrt{100}}} = -1$$

$$z_2 = \frac{x_1 - \mu}{\frac{\sigma}{\sqrt{n}}} = \frac{610 - 600}{\frac{600}{\sqrt{100}}} = 1$$

Hence, looking at the z-score table values, we can calculate the following:

$$\alpha = P(z < -1) + P(z > 1) = 0.1587 + 0.1587 = 0.3174 = 31.74\%$$

If we were going to assign the acceptance region to be between 580 and 620, we would reduce the Type 1 error to 4.56% (Calculation this statement is assigned as Exercise 5.6).

5.2 Testing Hypothesis Statements

Example 5.15 *(Sample Size's impact on Type 1 Error)* If we change the sample size in Example 5.14 and make it n = 400 then we change the z-value calculations:

$$z_1 = \frac{x_1 - \mu}{\frac{\sigma}{\sqrt{n}}} = \frac{590 - 600}{\frac{100}{\sqrt{400}}} = -2$$

$$z_2 = \frac{x_1 - \mu}{\frac{\sigma}{\sqrt{n}}} = \frac{610 - 600}{\frac{100}{20}} = 2$$

Noting that $P(z < -2) \simeq 0.0228$ indicating

$$\alpha = P(z < -2) + P(z > 2) = 2 * 0.0228 = 0.0456 = 4.56\%$$

Therefore, increasing the sample size by 4 times resulted in Error Type 1 to be reduced from 31.74% to 4.56%. Such increases in sample sizes typically reduce mistakes that can be made such as Type 1 error calculations. Excel and other programming languages can be used for easing the calculation of such values.

5.2.8.2 Errors of Type 2 Category

Given the meaning of Type 1 error, we will define and explain Type 2 error concept, and its impact on decision making in this section.

Probability Calculations of Type 2 Error. Type 2 error is the failure of rejecting the null hypothesis H_0 while it is false.

$$\beta = \text{Prob(type 2 error)} = \text{Prob}\left(\text{reject } H_0 \text{ while it is false}\right)$$

This error is also called **β-error**. Type 2 error requires another hypothesis that is alternative with another mean value.

Example 5.16 *(Type 2 Error)* In Example 5.14 we used n = 100 emails per minute used for Type 1 error and hypothesis testing for the rate x = 600. This sample is an estimation of the true population's mean value. Suppose that we reject the null hypothesis whenever the email distribution rate µ is greater than 610 emails or less than 590 emails per minute. Hence, we calculate the probability of a Type II error β for the values 610 and 590 and use this result to determine how the testing procedure would perform. For the same example, we used σ = 100 and µ = 600; Therefore, assuming that the data is approximately normal distributed (which should be checked for collected data sets), we have the same z-values as before in the example:

$$z_1 = -1 \text{ and } z_2 = 1$$

Therefore, noting that Prob($z < -1$) = 0.1587, probability of making a Type 2 Error is the following:

$$\beta = P(-1 < z < 1)$$
$$= 100\% - 2 * (15.87\%) = 68.26\%$$

This indicates 68.26% of all random samples would lead to rejection of the hypothesis $H_0: \mu = 600$ of the test procedure work when the sample size is 100 if we wish to reject H_0 within the critical region.

Example 5.17 *(Sample Size Change)* Suppose the sample size is increased from 100 to 200. We first need to calculate the z-values to calculate the Type 2 error:

$$z_1 = \frac{590 - 600}{\frac{200}{\sqrt{100}}} = -0.5 \text{ and } z_2 = \frac{610 - 600}{\frac{200}{\sqrt{100}}} = 0.5$$

By using the probabilities corresponding to these z- values by looking up the z-table values, we can calculate the Type 2 error:

$$\beta = P(-0.5 < z < 0.5) = 100\% - 2 * P(z < 0.5) = 1 - 2 * 0.3085 = 38.3\%$$

The probability of rejecting the hypothesis statement when it is false is 38.3%.

The **power of a statistical test** is defined to be the probability of rejecting the null hypothesis H_0 when the alternative hypothesis is true. The value of the power is 1–β.

5.2.8.3 Type 1 and Type 2 Error Related Considerations

- The errors Type I and Type II are related to each other. An increase in probability of one of the types of errors always results in a decrease of the other error's probability when the sample size is same.
- An increase in sample size reduces β-error in the case when α-error is held constant.
- If the null hypothesis is determined to be false, the value of β increases as the true value of the parameter approaches the value hypothesized in the null hypothesis. In the case when the difference between the true mean value and the hypothesized value increases, the value of β decreases.
- A Type 1 error (or significance level) of $\alpha = 0.05$ is widely used in hypothesis testing in practice. The choice of this value is based on the experience of the users and may not be appropriate for all situations.

5.2.8.4 Furthermore Advanced Hypothesis Testing Concepts

There are many factors that need to be incorporated into preparing, implementing, and finalizing hypothesis testing. Hypothesis testing is explained by using the normal distribution assumption however it can be also used for other statistical distributions that may require much more complicated mathematical calculations; therefore we do not cover such concepts in this book due to the overall purpose of the book, however readers can refer to other resource for more advance hypothesis testing concepts such as hypothesis testing on variance, t-distribution, and chi distribution related hypothesis testing analysis. Some of the steps to be taken during the application of hypothesis testing as a part of DMAIC approach are the following:

1. **Identification of Hypothesis Testing Areas.** Identification of the places where hypothesis testing can be applied is necessary prior to implementing any work. The distribution of the data (that relates to the defined problem) needs to be identified.
2. **Cleaning the data for processing.** Cleaning the data is an essential part before analyzing it and applying hypothesis testing. Depending on the nature of the data and its size, errors may be determined in a variety of ways but one way of determining errors in data is by using Excel. Other software packages can also be used.
3. **Identification of the null hypothesis (H_0).** Identify a null hypothesis H_0 in alignment with the defined problem.
4. **Identification of the alternative hypothesis (H_1).** Clearly state the alternative hypothesis H_1 based on the null hypothesis.
5. **Identify Test Statistic.** Identifying the test statistics that are in alignment with the Six Sigma team's expectations and goal to solve the problem on hand.
6. **H_0 rejection condition.** What conditions should be considered for rejection of the Null hypothesis?
7. **Numerical computations.** Carry out the necessary mathematical calculations by using identified information. Sampling quantities may be needed to be in alignment with the goal of the project.
8. **Do results make sense?** Identify if the mathematical computations in relation to the scope of the problem and the solution results make sense. If they don't then revisit the prior steps to make sense of the results based on the actual system under question and the data.
9. **Drawing Conclusions.** Deciding whether the null hypothesis should be rejected is the last step of the overall hypothesis testing process. Interpretation of the result needs to be based on the definition of the problem and the interpretation needs to be down to earth for a general audience to follow the results.

5.2.9 Displaying Measure Phase Outcomes

Identification of meaningful display mechanisms is as important as determining meaningful ways to display meaningful results from the statistical calculations. Your role as a Six Sigma team member is to be a person who evaluates the technical analysis information through the measurements and explains such information to stakeholders that may not have any technical background. The data collected is speaking with you a language and tells you the behavior of the system on hand and meaningful displays are the key elements of communicating such information to the stakeholders after you understand and interpret yourself after the analysis of the data. A challenging task during such display choices can be the recognition of which display needs to be used for the associated summary results. The displays need to have simple design features with no overwhelming design elements such as crowded data that cannot be followed at all, and they need to be readable and easy to explain in nature. What follows in this section is introducing and explaining to you some of the basic display structures that can be used during communication with non-technical and technical groups such as stakeholders of a cybersecurity event.

5.2.9.1 Histogram—Bar Chart

A histogram displays data in the segmented blocks as a result of created subsets within the data set; therefore, such a chart is particularly useful for displaying either grouped subsets of data that are demonstrated as "bars" or data values that need to be viewed in the form of bars. A bar chart is useful for both visual representation of a massive data set and displaying statistical calculations on the data if desired. The visual structure of a bar chart helps to understand the data distribution. We can also demonstrate bar charts with statistical measurements such as average and standard deviation of the data after calculation such quantities of the data. Such an approach can be helpful in determining average and standard deviation ranges of 2σ, 4σ, and 6σ from the average. Such graphs can also be used for demonstrating whether the data has normal or skewed distributions.

Example 5.18 Using the five levels Identify, Protect, Detect, Respond, and Recover that we covered, the following can be applied on the open issues:

- **Open Issue Priorities.** High, medium, and low
- **Open Issue Types.** Notable, significant, and critical

The following graph demonstrates the distribution of priority levels and types of open issues at the associated five levels. A way to measure a system's performance can be based on the leveling off techniques that can be used from Lean methodology. An example of such a methodology is the application of the Heijunka (production leveling) technique on the operations (Fig. 5.11).

5.2 Testing Hypothesis Statements

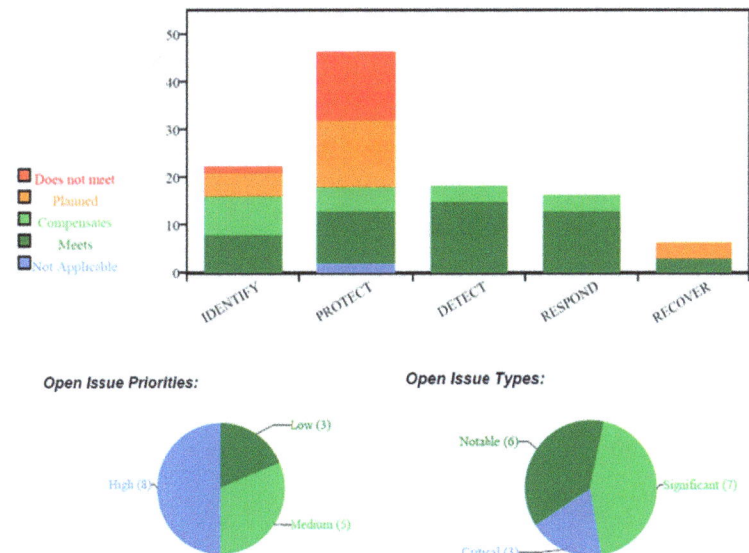

Fig. 5.11 An example of numbers of detected open issue priorities and types determined for the five stages Identify, Protect, Detect, Respond, and Recover

Another example is Example 4.3 where we covered vulnerability patching with certain production considerations. We can see below the quantities representing the different levels of production in a bar chart displayed side by side (Fig. 5.12).

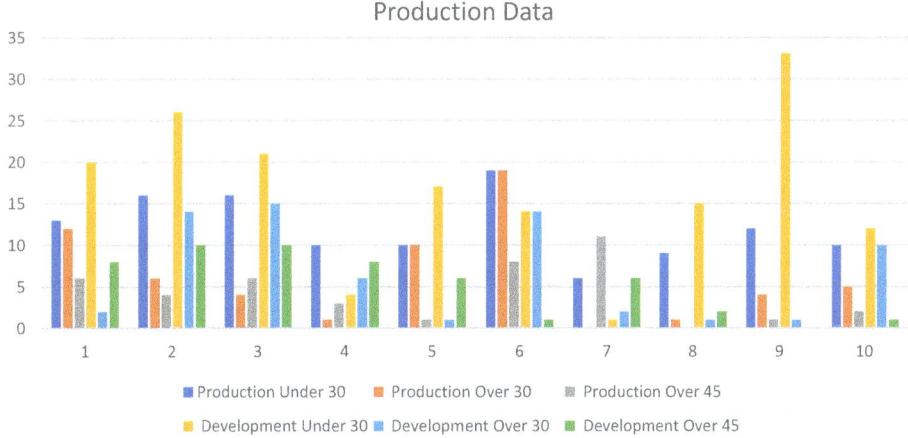

Example 5.19 *(Vulnerability Patching)* It is possible to represent the accumulated coverage of security tools by using accumulated data based on monthly information as follows:

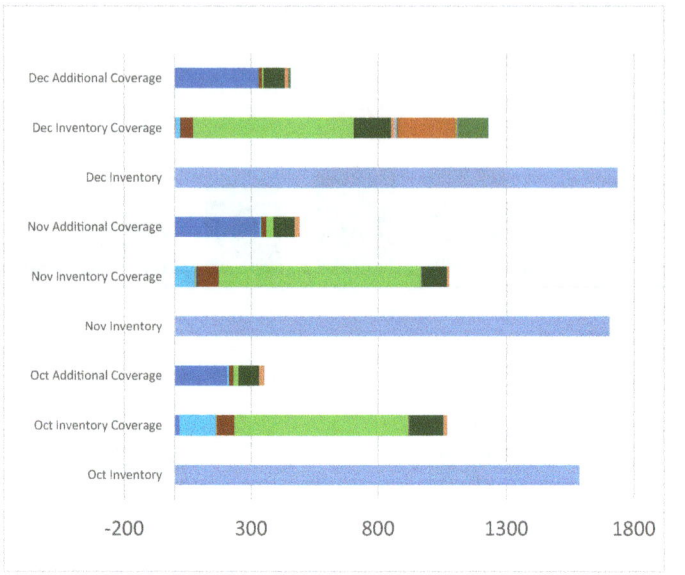

Fig. 5.12 An example of accumulated data as a bar graph reflecting security tool coverage

This graph displays effective inventory coverage for the number of cyber thread responses that occurred between October and December. A mix of the following security tools are used in the inventory:

- **AMP.** Antivirus and Malware Protection
- **PAC.** Privileged Access Control
- **SCCD.** Security Configuration Change Detection
- **SIEM.** Security Incident and Event Management
- **BAM.** Behavioral Anomaly Monitoring

5.2.9.2 Causes and Effects

The reasons for occurrences and the associated result can be outlined in a cause-and-effect diagram as a summary of possible causes with their effect on the output. Structuring a cause-and-effect diagram is helpful to determine the main categories of the problem as the problem is caused by them. A comprehensive cause-and-effect diagram's construction requires a comprehensive understanding of all possible causes of the outcome. After a comprehensive understanding, determining possible sub-causes under the main categories is also useful to understand the details. A cause-effect diagram is usually displayed in a tree structure with branches and sub-branches. It is also known as fish-bone diagram due its look like a fish with bones. The main categories of causes in the production of an item can be the following:

5.2 Testing Hypothesis Statements

- Methods used
- Labor (Manpower)
- Machines used
- Materials used
- Surrounding Environment
- Measurements used

Cause-and-effect diagrams' main categories may change as causes and the corresponding sub-causes may change. The following is an example of a cause-and-effect diagram designed for a network system possible failure.

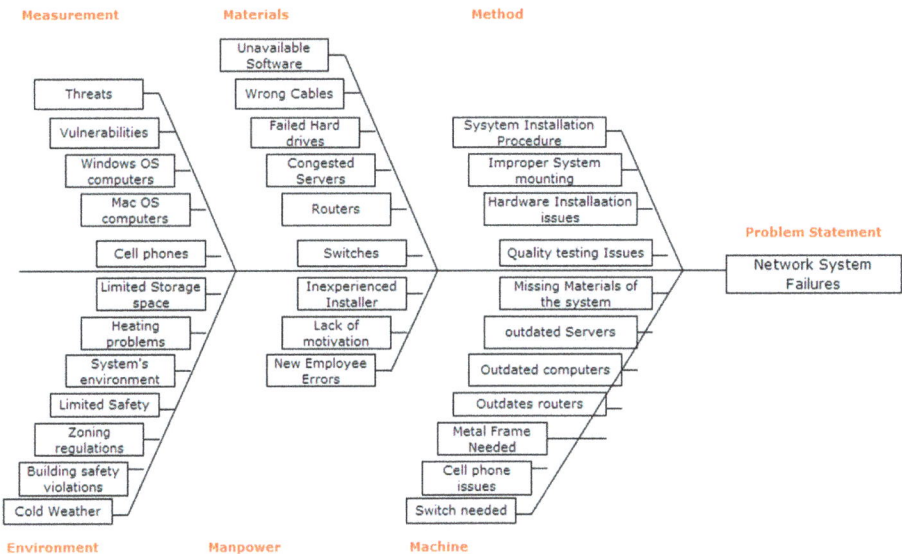

5.2.9.3 Run Chart/Time Series

A time series chart is a chart displaying the value of the data versus the output values. Time series charts are particularly useful for seeing the data visually. This visual representation is a transition from the data points collected as numbers to displaying them as points on the Cartesian coordinates.

5.2.9.4 Box Plot

Let $0 \leq m \leq 1$. 100mth percentile of a data set is a value that at least $100\,m\%$ of the data values are at or below this value and $100(1-m)\%$ of the data values are at or below this value.

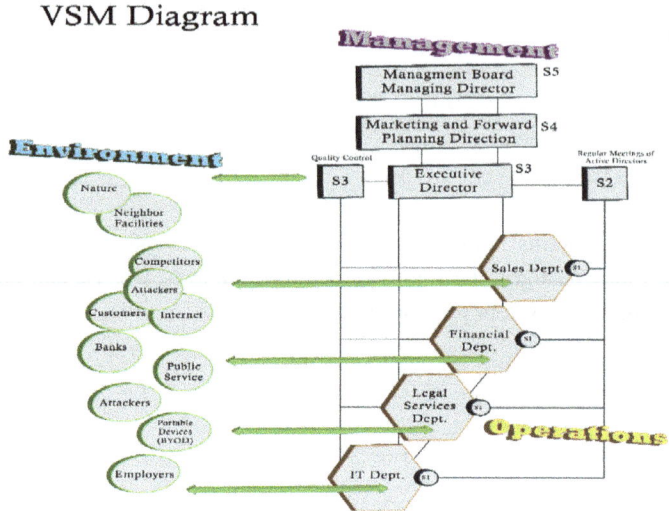

Fig. 5.13 Example of a VSM that makes a connection between operations, management, and environment

- First Quartile (Q1): The data values that lie at the bottom 25% of the data set.
- Third Quartile (Q2): The data values that lie between 25 and 50% of the data set.
- Third Quartile (Q3): The data values that lie between the median and 75% of the data fall in this category.
- Fourth Quartile (Q4): The data values that lie between 75% and maximum of the data fall in this category.
- Outliers: Any statistically extreme value that cannot be classified to be in the above-mentioned four quartiles. The determination of the outlier depends on the data set values and the scope of the project.

5.2.9.5 Value Stream Mapping (Data Entry Phase)
See (Fig. 5.13).

5.2.10 Takt Time (TT)

Takt time is particularly useful to measure time-dependent changes in a system Takt times of pre- and post-improvement stages of a system can be compared for rate of improvement purposes. Takt time is the time it takes for stakeholder-expected on-time output delivery by the process starting with the input's processing. It is also an indicator of processing capability from a time perspective. It is helpful to quantify the production rate for stakeholder reporting. The following are essential to know for calculating the takt time:

5.2 Testing Hypothesis Statements

- **Hours at Work (H).** The time that machines/humans work after taking out all the time that they do not work during production.
- **Stakeholder Expected Demand (D).** Expected amount of production/service by the stakeholder.
- **Period (t).** The time period used for the amount of work to be completed.

The formula of takt time is the following:

$$TT = \frac{H}{D}$$

Using the takt time, the number of operators needed to complete a task can be also calculated:

$$\# \text{ of operators} = \frac{Takt\ time}{t}$$

Example 5.20 The effectiveness of the responses to threats can be identified through takt time calculations to identify the number of servers or additional tool coverage needed to respond to threats at a certain rate. Given that the intend is to calculate the ratio of total available production time to average system or customer demand, such calculations can be easily carried out for cybersecurity applications.

Suppose a cyber security system needs to process 12,000 operations per year. This would require determining the number of servers to be placed with the associated tool coverage for system's threat management for fulfilling the expected operations under the following conditions of servers:

- Working 24 h per day
- Shut down four times with each time lasting 10 min as a break
- Have downtimes of 50 min

We can determine the takt time by using this information. Suppose the total cycle time of the servers is expected to be 15 min, we can also determine the number of servers needed. We first convert the information into work hours to have the same unit calculations. The expected demand is the responding of 12,000 operations per year hence d = 12,000 operations per year. If the operation handling occurs throughout 365 days (this can change depending on the organization,) we have

$$\frac{12000}{1 year} = \frac{12000\ operations}{(365\ days) * (24\ hours)} = 1.37\ operations\ per\ hour$$

Hence the operations response is expected to occur about 2 operations per hour.

- Hours at work. The servers work 24 h per day, have four breaks of 10 min, and a downtime of 50 min. Therefore, the amount that the servers spend within the system is

$$h = 24\,hours - [4*(10) + 50\,minutes] = 22.5\,hours$$

- Period.

$$t = 15\,minutes$$

- Takt time.

$$Takt\,time = \frac{22.5}{1.37} = 16.423\,hours\,per\,operation$$

- Number of servers. The corresponding number of servers needed is the following:

$$\#\,of\,servers = \frac{16.423}{0.25\,hours} = 65.693$$

We need this number of up to 66 servers because we need more server power than 65.693 to be able to handle the operations. It is important to note that these calculations need to be compared with the data collected and ensure that this number of servers would be sufficient to handle the operations based on the peak times and data occurrences. The number of servers also provides information about threat management and the associated analysis.

5.2.11 Overall System Effectiveness

The effectiveness of a system can be measured in several ways. One way to measure it is through the evaluation of the method used for fulfilling a task by using good quality products or service quantities over the entire system's quantities. The effectiveness of equipment or software within a system can be measured based on the measured instances while the service effectiveness can depend on the possible measurements such as the number of services provided per unit time. Figure 5.14 explains the overall system performance based on the measurements of good and defected donuts produced. The ratio of good donuts to all donuts is a measure of system's effectiveness during production. Improvement of such a production system would yield to more good quality donuts that would increase the overall methodology's effectiveness.

Measuring production numbers and improvement outcomes due to reduced waste is meaningful by using OEE. For instance, if we want to determine the rate of failures caused by a server within a system to access the system, we can measure it by using OEE

5.2 Testing Hypothesis Statements

Overall Methodology Effectiveness

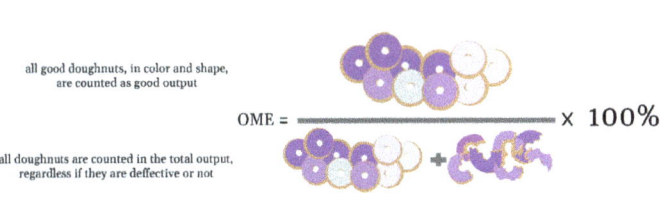

$$OME = \frac{\text{good output received using the particular method}}{\text{total output}} \times 100\%$$

Fig. 5.14 Formulas for OEE/OME to calculate values that relate to overall methodology effectiveness for a doughnut example

by carrying out the corresponding calculations. This method can report the effectiveness of the server states and failures with their impact on the outcomes to evaluate the method used for accessing the network. Examples of waste that can be measured include but are not limited to the following:

- **Systematic Setup.** Setup of systematic parts may cause challenges during their setup to start operating and such time-related issues can be improved by changing the methodology of the set up. It is known as the downtime for the machine during operational times.
- **Human Mistakes.** Mistakes made by the users of the systems can cause a variety of wastes that are the results of several mistakes. The effectiveness of the system can be improved by measuring and focusing on correction of such mistakes.
- **Breakdown waste.** Breakdowns of machines can occur in a system and such breakdowns can cause waste easily. This is also considered to be a downtime of the system during operational period.
- **Pace-effecting Waste.** Several machines, personnel, and systematic components may cause the system to slow down and impact the pace of the overall system. Such issues would impact the operational efficiency of the system and cause the system to not have the best performance.
- **Design Waste.** There can be several wastes due to the design of the system in place and such waste can be eliminated by replacing the inefficiencies with efficient methods. Such waste impacts the quality of the systematic processing.

Fig. 5.15 Speed, quality, downtime, and VA time distribution of total production time as a pie chart

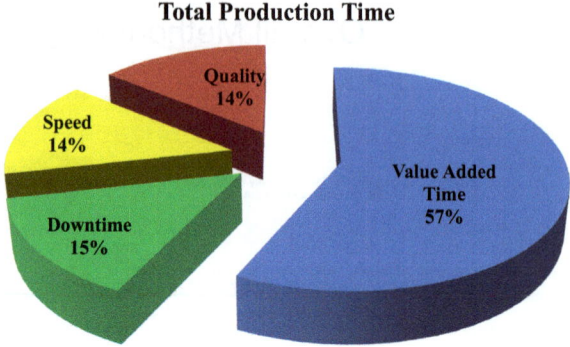

- **Defective waste.** Some of the processing work completed by an entity of a system may not only cause challenges with running the system but also cause other system components/entities to be impacted by the issues produced by the entity. Such a defective entity can cause defective processing and waste during processing (Fig. 5.15).

5.2.12 Measuring Systematic Performance

Systematic performance measurements can be implemented with the typical "per unit" approach. The definition of unit depends on the quantities considered within the context of systematic considerations. In cybersecurity applications, it is possible to deal with numbers within the range of the first 100 numbers while also dealing with numbers in millions. Our focus in this section is going to be on methods to scale the performance measures of a process to measure the success of the process. By using the term defect, it is referred to as anything that may distract the system from the ordinary functional nature of it and causes cyber-threat. Correcting via rework is considered to be the ability to determine the cause of the cyber-threat and eliminate its effectiveness in the system while failure to correct after rework is considered to not find a possible resolution for the cyber-threat. For instance, an email can be considered defective if it contains a virus or a threat is not handled in the way it is expected. Identifying the types of emails as threats and placing them in junk email is correction after rework. The performance measures that we will cover include the following:

- Defects in a Million Opportunity (DMO)
- Defects identification per Unit (DPU)
- CTQ Defective Parts in a Million (CDPM)
- Process Sigma
- Overall Equipment Effectiveness (OEE).

5.2 Testing Hypothesis Statements

5.2.12.1 Defects in a Million Opportunity (DMO)

Identification of defects in a million units represents the number of possible defective items in a million possible occurrences. For instance, this approach can be particularly useful in simulating attack calculations. Let

- Td: Total number of defective items (i.e., threats that turned out to be causing problems in the system)
- n: Number of threats processed
- d: Expected number of defective entities in production

The formula corresponding to DMO is

$$DMO = \frac{Td}{n*d} * 10^6$$

Example 5.21 Suppose the following are observed in a system:

- Number of items inspected = 1000
- Total number of defective threats found = 100
- Number of opportunities per unit = 5

Therefore

$$DMO = \frac{100}{1000*5} * 10^6 = 20000$$

defects are detected per million items produced.

5.2.12.2 Defect Identification Per Unit (DPU)

Identification of defect rate based on the basic unit of the entity is calculated by using defect identification per unit formula:

$$DPU = \frac{d}{n}$$

The output of this formula is the average number of issues over the total number of items processed.

Example 5.22 There are 50 threats detected during the processing of 200 instances. After reprocessing these threats, 40 of them are corrected. We want to calculate the DPU values for pre- and post- correction periods when there are 1000 threats occurred.

After the first phase of processing, the average number of defects is

$$DPU_1 = \frac{50}{200} * 1000 = 250$$

threats per thousand threats occurred. After rework, 40 of such threats are resolved, therefore there are 10 remaining in 160 instances. Hence, the average number of threats after rework can be recalculated as follows:

$$DPU_2 = \frac{10}{160} * 1000 = 62.5 \cong 63$$

threats per thousand instances.

5.2.12.3 CTQ Defective Parts in a Million (CDPM)

Customer approved defective items can be different from the typical defective items as the customer's approach to the definition of defect may be different from other persons. Hence, CTW defective parts in a million allows us to measure the defects that respect the customer's Critical to Quality conditions per million (CDPM). This allows us to determine the average number of defective items in a million units where an item is defined as defective if it is not approved by the customer to be normal. To determine CDPM, we define.

- n: Number of items produced
- c: Customer approved defective items

Hence

$$CDPM = \frac{c}{n} * 10^6$$

Example 5.23 Recall Example 5.22 of the 40 threats eliminated, 30 of them were identified as not considered as threats by the stakeholder. We want to calculate CDPM and DPU of this production.

Out of the 40 threats, noting that 30 threats are not aligned with CTQ, only 10 of them agreed with what the customer wanted to see eliminated. This means, there were only 10 threads that were considered as threats among the 20 potential threats (due to 50–30 = 20). Therefore, from the customer's perspective, there were 20 possible threats with 10 of them corrected. Therefore

$$CDPM = \frac{10}{200} * 10^6 = 50000$$

items in a million threats occur. From defect per unit perspective, out of 50 threats, 40 were corrected therefore

$$DPU = \frac{40}{200} = 0.2 \text{threats per unit}$$

5.3 General Software and Online Resources for Six Sigma Analysis

This section is particularly designed for software packages that can be used for measurements and displaying statistical results and summaries of these measurements. We introduce these tools that can be used for data analysis in this section because it makes sense to know where the data should be stored in order to be able to use the software in the Analyze section. Many professionals tend to use the software packages that they are comfortable with from the beginning of a six-sigma project because of possibly not necessarily knowing the strengths and weaknesses of the other software packages along with the fact that they are "more comfortable" with the software packages they learned early during their education or career. It could be also because the company that they work for may require them to use the corresponding software. Regardless of the argument of software choice, in this section, the data sources that can be accepted by the corresponding software packages will be covered in this section. This information on the software packages is particularly important for collecting and storing measurement data along with their brief use for data analysis. Details on the ways to utilize these software packages for data analysis purposes will be explained in the *Analyze* chapter of DMAIC for six sigma project purposes. The software packages to be covered in this section include the following:

- C
- C++
- Java
- IBM SPSS
- Matlab
- Minitab
- Microsoft Excel
  - General Functions
  - Visual Basic for Applications (VBA)
  - Add-ins
- Microsoft Visio
- Palisade Products
- Python
- R and R-Studio
- SAS and SAS-Studio

5.3.1 C

C programming language developed in Bell Labs around 1972 is still one of the most popular programming languages despite its low-level capabilities. C is very flexible and versatile that allows high-level control over the codes written for microcontrollers and operating systems. One of the advantages of C is that it is the foundational language for object-oriented languages that are popularly used, such as C++, C#, and Java. There are two basic types of files that can be used within C programming:

- Text files. Text files are **.txt** files and they can easily be created by using any simple text editors such as Notepad.
- Binary files. Binary files are mostly **.bin** files in which the data appears in the binary form (0's and 1's) instead of text format.

C programming language might be a low-level desired programming language for six-sigma project applications since there are more advanced programming languages for statistical analysis and packages that can be used for statistical applications.

5.3.2 C++

C++ programming language was found shortly after as a part of a PhD dissertation of a doctorate student. The " + +" is the variable increment in the language indicating that it is an update of C. Objects are added to the C language for practical uses to derive the C++ language which made it an object-oriented language. C++ is one of the powerful languages and yet requires plenty of work to be done for six-sigma projects therefore we don't recommend using it unless C++ is your primary programming language, and you are extremely comfortable with it in many ways. Like the C programming language, text and binary files can be read by the C++ language however it doesn't mean the information cannot be read from other files. Any information can be transformed into C++; however, the meaning of the transformed data needs to be interpreted as a part of the code written by the programmer. One of the strengths of C++ is its ability to take in large data sets and be able to analyze them. There are certain benefits of using C++ in big data applications:

- C++ Enhances Processing Speed. Large terabyte or petabyte data sets need to be processed reasonably quickly when complex algorithms for machine learning are involved. C++ is the language where more than 1 GB of data can be processed in a second. Further, you can retrain and apply predictive analytics in real-time, maintain consistency of the system of recordings. Data scientists often use C++ to write big data frameworks and libraries

- Enables System Programming. There are many deep learning algorithms that require implementation in C++.
- C++ applications typically need less capacity and electric power than virtual machine languages therefore they can be considered cost effective and efficient.

5.3.3 Java

Java, developed in the 1990s, is an object-oriented programming language with objects defined in certain classes. It is one of the most popular programming languages that is used by program developers in the World today. Java code can run on all platforms that support Java without the need for compilation. It is possible to open and read any file in the text or binary format, however the information needs to be converted into meaningful outcomes just like it is stated for C++. Unless Java is your primary programming language that you would need to read and interpret data, we don't recommend using it for statistical data analysis and displaying figures. One of the strengths of Java, similar to C++, is its capability to take in large data sets and be able to analyze them.

5.3.4 IBM SPSS

IBM SPSS (Statistical Product & Service Solutions) Statistics is a popular program for statistical analysis in social sciences. Like Minitab, SPSS has its user-friendly data entry platform similar to Excel and does not require coding with a point-and-click platform. SPSS can be integrated with open-source platforms such as Python and R programming languages as extensions of the software. It is possible to analyze large and complex data sets in SPSS. One important feature of SPSS that can make it distinguished from other software is data cleaning; faulty values and missing data can be identified easily by using SPSS. Storage of data in your local drive rather than cloud makes it a much more secure way to keep the data when SPSS is used. The data can be transferred into SPSS by using multiple formats:

- Excel spreadsheet
- Text file (.txt,.dat,.csv)
- Database

The transferred data is then stored as an SPSS file with the extension.sav. SPSS can be used with Windows or Mac operating systems. The following image displays SPSS' software environment and the analysis of the data along with the displayed data on the screen.

5.3.5 Matlab

MATLAB is one of the strongest programming languages with plenty of features and apps to be utilized. It has many built in Apps (such as optimization, curve fitting, Signal Analyzer etc.) that help with specific goals of users. There are many file types that you can store your data and be able to call the file in MATLAB's environment for analysis. Data can be imported into MATLAB in many different formats, and it can also be linked to other programming languages.

5.3.6 Minitab

Minitab is a software package that is well known for its statistical analysis capabilities. There are many functions and graphing tools that you can use in Minitab without any script used. The program works mainly as "click-and-choose" for attaining solutions. The software has an Excel-like environment which allows users to enter their data in this environment. The output data obtained from the program can be also easily transported to Microsoft Excel, Word etc. for practical use. The Navigator section of the software allows the users to keep track of their solutions and edit them. The number of cells that Minitab can import is limited by your available memory address space. When you open a file with Minitab, the file's contents are copied into the project. Therefore, if you change the data or graph contents within a project, you do not affect the original file. The supported file types include the following:

- Excel file (XLS, XLSX, XML)
- Text file (CSV, TXT, DAT)
- Minitab file (MPJ, MTW, MGF)

For instance, the following image in Minitab's environment is bootstrapping of 100 randomly generated data points by using Normal distribution. The data in the following figure is generated by using Random Data generator of Minitab along with the Resampling component within the software.

5.3.7 Microsoft Excel

Majority of the companies in the World work with Microsoft Excel for storing and analyzing their data unless they have enormous amount of data to be collected and stored. Excel is particularly popular among the manufacturers to store and analyze data after it is collected within the same environment. It is challenging for Excel to handle data analysis; therefore, the software can crash when there is an extreme number of data points are stored and attempted to be analyzed. Excel has very valuable features such as general mathematical functions, Visual Basic for Applications programming (i.e., VBA), and add-ins, the plug-ins to Excel that can be used for certain purposes in a simplistic fashion. In this section, we will cover the basics of these features of Excel and how they relate to the Measure phase.

5.3.7.1 General Functions

Suppose you start storing your data in Excel and you want to implement basic statistics such as mean, standard deviation, median, and mode values, then you can use Excel functions that can be typed in the cells in Excel. In an Excel cell once you start with the equal sign (i.e., $=$) then the rest follows with your choice of function that you want to type in. Excel helps you with the choice if you are familiar with the functions. Many daily users of Excel don't necessarily know this feature of Excel and therefore don't get to take advantage of this simple and useful feature of Excel. Once you start typing your choice of function you then the function is shown automatically by Excel (in the image below) and you can click on the "tab" key on the keyboard to automatically choose the function instead of typing it all. Then the rest follows to simply choose the range of the cells in Excel that you would like to apply the function and hit enter.

There are many more useful functions that you can find in Excel if you would like to explore more. Microsoft's support webpage outlines all the functions that you can find in Excel. You can also use pivot tables (under the Insert tab) that can help you to group and categorize large data sets to extract meaningful outcomes from them. For instance, the following image displays the percentage distribution of a data set that contains data between 0 and 0.7. The pivot table output is the percentage of the data in the entire data set.

Row Label	Count of 10
0-0.1	20.00%
0.1-0.2	10.00%
0.2-0.3	10.00%
0.3-0.4	20.00%
0.5-0.6	20.00%
0.6-0.7	20.00%
Grand Total	100.00%

There are books written on Excel functions and their use therefore we are not going to cover most of these functions in this book; our goal is to only give you an idea of the resources available and indicate the availability of the choice.

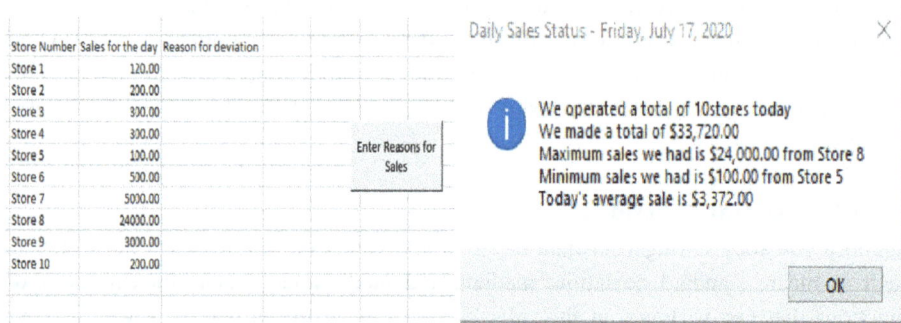

Fig. 5.16 An example of VBA's output with the button feature and automated response output

5.3.7.2 Visual Basic for Applications (VBA)

VBA is the programming language for Microsoft Office products such as Excel, Word etc. You can create Macros in Excel that will help you to automate analysis solutions that you would like to implement. You can record macros for simple applications in Excel to initiate a solution and then edit it. "Macros" is available under the "View" tab in Excel. For instance, a program can be designed for searching on the sales of your interest in an Excel spreadsheet by writing the associated code. This program interfaces with the Excel cells to complete tasks in the way you design solutions.

There are very strong and useful features that you can use in the Macro setting of Excel. You can generate Macros that pop-up displays for the users to enter choices for search or create buttons for users to click on a button to accomplish a task on a data set. For instance, Fig. 5.16 demonstrates a button that can be used for entering the reasons for sales and the corresponding calculations on total, maximum, minimum, and average sale amounts based on the entered data in Excel.

5.3.7.3 Add-ins

Excel Add-ins are custom-made Excel solutions that target a specific purpose to be fulfilled and may need to be purchased depending on the applications that you would like to use it for. You can think of Excel Add-ins as a set of Macros that are designed and packaged to serve a certain goal. You can view the Add-ins available in your version of Excel by looking at Options as a part of Add-ins section. For instance, as can be seen in the image below that @Risk by Palisade, Analysis ToolPak, and Solver add-ins are some of the add-ins you can find online in applications. The Analysis ToolPak is an Excel add-in program that provides data analysis tools for financial, statistical, and engineering data analysis. Some of the applications of Analysis ToolPak include calculations of correlation, covariance, and descriptive statistics.

5.3.8 Microsoft Visio

Microsoft Visio, apart from the professional statistical software packages and data entry environments that we have stated so far, is a software used for drawing flowcharts, diagrams, org charts, floor plans, engineering designs, and many more. It contains modern shapes and templates with the familiar Microsoft Office experience. You can map an IT network, build an organizational chart, or document a business process. In addition, it is possible to use it with Microsoft 365.

5.3.9 Palisade Products—@Risk, Evolver, NeuralTools, PrecisionTree, StatTools, TopRank

Palisade provides several Excel add-ins including @Risk, Evolver, NeuralTools, Precision Tree,StatTools and TopRank. These add-ins use Excel's data environment; therefore, any data that can be imported into Excel would allow the usage of the listed add-ins.

5.3.10 Python

Python is a free programming language, and it is one of the most user-friendly object-oriented programming languages that can be used for data analysis. It is easy to use because of the structure of its script and the design of the language. You can create and use classes and objects as a part of the analysis. These packages are particularly useful for reading and analyzing data. For instance, machine learning has its own classes and objects used in applications.

A file format is a standard way in which information is encoded for storage in a file. Measured data information stored in many different file formats including.

1. Comma-separated values
2. Excel (XLSX)
3. ZIP
4. Plain Text (txt)
5. JSON
6. XML
7. HTML
8. Images
9. Hierarchical Data Format
10. PDF
11. DOCX
12. MP3
13. MP4

can be read by using Python. The data needs to be stored carefully by using the corresponding guidelines and the format.

5.3.11 R and R-Studio

R is a programming language and environment for statistical computing and graphics. Statistical analysis and modeling such as linear and nonlinear modelling, classical statistical tests, time-series analysis, classification, clustering, etc. is possible by using R as well as graphical techniques. R is an open source programming language under the Free Software Foundation's GNU General Public License in source code form, and compiles and runs on a wide variety of UNIX platforms and similar systems (including FreeBSD and Linux), Windows and MacOS.

R allows users to add additional functionality by defining new functions and it can be linked to codes written in programming languages such as C, C++, and Fortran for computationally intense tasks. C code can be used to manipulate R objects directly.

R is recognized as an environment in which statistical techniques are implemented. R can be extended (easily) via packages. There are about eight packages supplied with the R distribution and many more are available through the CRAN family of internet sites covering a very wide range of modern statistics. R has its own LaTeX-like documentation format, which is used to supply comprehensive documentation, both on-line in several formats and in hardcopy (Ref. https://www.r-project.org/about.html).

A simple text file is the easiest form of data to import into R therefore collecting data in a text format or a format that can be converted into text format is a good method for storing collected data. The primary function to import from a text file is scan, and this underlies most of the more convenient functions discussed in Spreadsheet-like data.

If you are receiving the data from the client, the data can be shared with you in some proprietary binary format, for example, an Excel spreadsheet' or 'an SPSS file'. Using the originating application to export the data as a text file is often the easiest way to handle it. However, this is not always possible, and importing from other statistical systems discusses what facilities are available to access such files directly from R. For instance, this is possible for Excel spreadsheets.

Data can also be stored in a binary form for compactness and speed of access. An example of such modality is imaging data, which is normally stored as a stream of bytes as represented in memory, possibly preceded by a header. Such data can also be imported into R.

A database management system (DBMS) for much larger databases is the best way to handle the data. The DBMS can be used to extract a plain file, or the extraction operation can be done directly from an R package for many DBMS. We refer to (Ref. https://cran.r-project.org/doc/manuals/r-release/R-data.html#Imports) for more details on importing data into R.

RStudio is an integrated development environment for R that is available as RStudio Desktop as a regular desktop application and RStudio Server that runs on a remote server and allows accessing RStudio using a web browser. R-studio has many more features including.

- RStudio (open source)
- RStudio server (open source)
- Shiny Server (open source)
- R packages (open source)
- RStudio Cloud (hosted services)
- Shinnyapps.io (hosted services)
- RStudio Team (Professional—Enterprise ready)
- RStudio Server Pro (Professional—Enterprise ready)
- RStudio Connect (Professional—Enterprise ready)
- RStudio Package Manager (Professional—Enterprise ready)

Recently R and Python are combined in RStudio for using these two languages within the same environment. Python files can be opened in RStudio and RStudio allows users to view statistical analysis results automatically within the coding environment. The underlying method of merging these two programming languages arose from the use of reticulate and the common interest in C++. Therefore, Reticulate embeds a Python session within your R session, enabling seamless, high-performance interoperability.

5.3.12 SAS and SAS-Studio

SAS©, known as Statistical Analysis System, is developed for data management and advanced statistics in 1976. It is written in C programming language and can be used in Windows, IBM mainframe, Unix, Linux, and Open VMS Alpha. SAS program can be used by writing SAS language or point-and-click interface. Data collected in the following formats can be imported into SAS:

- Delimited files, such as CSV, TXT, TSV, DLM.
- dBASE V, IV, III+, and III (DBF).
- Stata files (DTA).
- Microsoft Excel files (XLS, XLSX). To import XLSX files, you must license and install SAS/ACCESS to PC Files.
- JMP files.
- Paradox DB files.
- SPSS files.
- Lotus 1-2-3 files from Releases 2, 3, 4, or 5.

SAS Studio is the University Edition of SAS. Many people program in SAS by using an application on their PC desktop or SAS server. SAS Studio is different because it is a tool that you can use to write and run SAS code through your web browser.

Chapter 5 Exercises

Exercise 5.1 What is the major difference between type 1 and 2 errors. Give an example of these types of errors in cybersecurity applications. Please explain your response.

Exercise 5.2 Give an example of each one of weaknesses and strengths of hypothesis testing in cybersecurity applications. Please explain your responses.

Exercise 5.3 What may be the most important aspect of critical measurement in cybersecurity applications. Please explain your response.

Exercise 5.4 What are the factors that impact the overall system effectiveness of a cybersecurity system? Please explain your response briefly.

Exercise 5.5 What can be the critical elements of preparing a survey for distribution to professionals for a cybersecurity six sigma project? Give an example and explain your choices and reasoning to choose such survey design elements.

Exercise 5.6 If we change the acceptance region in Example 5.14 to be the values between 480 and 720, how does the Type 1 error change?

Biobliography

1. C programming language. https://computer.howstuffworks.com/benefits-learning-c-programming.htm
2. SPSS© by IBM. https://www.ibm.com/products/spss-statistics
3. Minitab©. https://www.minitab.com/en-us/
4. Microsoft Excel©. https://support.microsoft.com/en-us/office/excel-functions-alphabetical-
5. Palisade©. https://www.palisade.com/risk/
6. Microsoft Visio©. https://www.microsoft.com/en-us/microsoft-365/visio/flowchart-software
7. Python©. https://www.python.org/
8. R-Studio©. https://rstudio.github.io/reticulate/
9. SAS© by IBM. https://support.sas.com/software/products/university-edition/faq/SASStudio_whatis.htm
10. MATLAB© by Mathworks. https://www.mathworks.com/products/matlab.html
11. Java by Oracle. https://docs.oracle.com/javase/8/docs/technotes/guides/language/

12. Dale, N. B., Weems, C. (2004). *Programming in C++*. Jones & Bartlett Learning
13. Tokgöz, E. (2024). Quality & lean six sigma applications for industrial engineers. Springer Nature, Switzerland. https://link.springer.com/book/9783031557392
14. Tokgöz, E. (2024). Quality and lean six sigma for engineering technicians, synthesis. Lectures on engineering, science, and technology. Springer Cham 978-3-031-44033-5. https://link.springer.com/book/9783031440328
15. Tokgöz, E. (2025). Artificial bee colony optimization techniques' utilization for intrusion detection systems' analysis. In *4th IEEE International Conference on AI in Cybersecurity (ICAIC) proceedings*. https://ieeexplore.ieee.org/stamp/stamp.jsp?tp=&arnumber=10848880
16. ISO/IEC 27001 (2022) Information security, cybersecurity and privacy protection — Information security management systems — Requirements, Edition 3 2022. https://www.iso.org/standard/27001

Analyze the Measurements and System

6

In this chapter some of the tools that can be used for analysis of the measured information for the Six Sigma project will be covered. This analysis is possible by using the probabilistic results we covered in earlier chapters, sketching figures for the collected data, meaningful explanations of the graphs, and some of the statistical concepts that we will cover in this chapter. Correlation and regression analysis of the system performance, determining root causes and their effects, the use of technology for analysis purposes are some of the important concepts to be covered in this chapter. Once measurements with the associated metrics are identified for the scope of the processes under review, they may be analyzed as follows:

1. Implement a comparative analysis of the procedure outcome with the process goals.
2. Decide whether the metrics used are aligned with expected procedures.
3. Identify any measures that are not applicable to any metric and assign the activity underlying the measure as a potential NVA.
4. Analysis should result in measurable outcomes with VA and NVA activities determined and clearly shown on the process map after determining weaknesses of the process.

Risk hierarchy and CTQ are directly correlated with each other due to their relationship. In such a case. A desktop attack could be related to one of the following:

- Web-based malware used
- Worms embedded
- Phishing
- End user hacking.

Regarding cybersecurity related incidents, the following are essential for the analysis of the associated data:

- Collateral damage
- Analysis of harmed integrity
- Analysis of harmed confidentiality
- Insider threat
  - Error analysis
  - Malicious activity analysis
- Vulnerability exploits
  - Known vulnerability analysis
  - Zero-day attack analysis.

It will be ideal to use existing cyberthreat and cybersecurity process documentation to explain what, where, by who, when, how, and by how much improvement in the process to mitigate risk. Using a Kaizen event results resolve issues for larger scale problems that can be worked on as Six Sigma projects is one way to determine a Six Sigma project. A Kaizen event completed during a short period can become a six-sigma project for a longer-term improvement and extensive data analysis may be needed. Example approaches to waste identification are listed by category below.

- Underproduction:
  - Review email to find examples of phishing and malware delivery.
  - Look at patch history to see where security vulnerabilities exist on systems.
  - Determine how much value it adds to the current operations.
- Overproduction:
  - Review the support tickets to identify situations where security issues impacted productivity. This would have both the qualitative and quantitative data for analysis that can be an extensive Six Sigma project.
  - Survey users and ask for examples of security messages that made them pause in their work that turned out to be false alarms.
- Overproduction:
  - Sketch the locations of determined improvements on the operation map by using icons.

6.1 Five W and How Questioning

The components of Five W (5W) and How questioning approach require us to ask as five questions that relate to the analysis of the Six Sigma problem are on hand [22]. The 5Ws are the following:

6.1 Five W and How Questioning

- What
- Who
- When
- Where
- Why

These questions need to be asked in a smart fashion that directly points to either the reason for the problem occurring or the solution method investigation. In addition, we need to answer a question that uses "How".

The following steps can be followed to be able to apply this technique:

1. Problem identification.
2. Answer the five W's starting with the W that needs to be answered first that is crucial to solve the problem.
3. Answer all the Ws and "how" in a specific order considering their level of importance to the question.
4. Start constructing a cause-effect diagram.

Example *(Incident Time to Close)*. After measuring the data and determining the numerical results with visuals, the following sense making 5W and How questions arise:

- **What** causes different levels of priorities to have different correlations?
- **Where** is the best place to apply for the improvement? It initially appears to be the low-level priority based on the issues seen.
- **Who** can be the best person to determine the causes of variations in different levels of priorities?
- **When** can the analysis of variation be observed in the future?
- **Why** does the correlational relationship between analyze, mitigate, contain, post-mortem, and full recovery stages vary too much for different levels of priorities and the majority of the issues arise from low level prioritized instances?
- **How** can priority levels be adjusted and unresolved issues be minimized?

Example *(Vulnerability Patching)*. One can ask the following 5W and How questions for the *Vulnerability Patching* example.

For variational analysis,

- **What** is causing too much variation throughout the months between service levels?
- **Where** is the source of variational causes?
- **Why** is variation occurring among different levels of responses?
- **Who** oversees deciding responses?

- **When** can the minimum response time be improved?
- **How** are the variations that are occurring handled?

For analysis towards improvement,

- **What** needs to be done for improvement?
- **Who** oversees approving improvements?
- **Why** do we need improvements?
- **Where** are the primary improvements needed?
- **When** can the improvements be the most powerful to apply?
- **How** can the improvements be implemented?

6.2 Batch Means Method

The analysis of large volume data points (e.g., 13,500 within three months) can be sometimes challenging and cumbersome by using basic known analysis techniques in which case the batch means method can be useful. This method requires dividing the dataset into batches of data points for analysis by using statistics. The average of the data points in each batch needs to be calculated along with correlation, covariance, autocovariance, and autocorrelation. For demonstrating this method, we take a small sample data set of the full-recovery information from the low priority cases in the *Incident Time to Close* data set. If we try to graph the data as is based on instances, the large data points would overwhelm the graph and wouldn't show a good representation. Figure 6.1 displays the averages of 12 batches with each consisting of 30 data points which only contain the cases where no issues occurred.

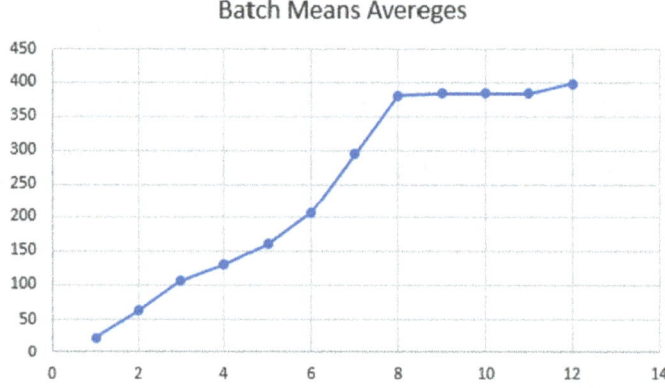

Fig. 6.1 Application of the averages of batch data collected with each point representing the average of 12 subsets existing within the data set

6.2 Batch Means Method

Fig. 6.2 Batch means graph for issues occurred with low priorities

Figure 6.2 displays the issues that occurred for the same dataset which is separated from the 12 batches noting that its average 40,300 s is a large number that cannot be added as a data point on the Batch Means Average graph. The Batch Means Graph indicates the stable averages from batch 8 to 12 which is around 380. This possibly means there are about 4 * 35 = 140 data points which are close to 380 noting that the graphs are attained after sorting the data. We also see in the linear trend of increase in this data set (when sorted) with the possible minimums and maximums in the data for both cases when issues and no issues occur.

The batches need to be big enough so that the batch averages would not be highly correlated. When the data is time dependent, there will be correlation for the collected data because the last item of the previous batch would always be correlated with the first item of the next batch. However, noting that this transition from batch to batch can be small, it can be ignored. The details of the Batch Means analysis require more advanced statistics with the following formulas' use:

- $cov(X_1, X_2)$ representing the covariance between X_1 and X_2.
- $cor(X_1, X_2)$ representing the correlation between X_1 and X_2.
- σ_{X_i} representing the standard deviation of the random variable X_i.

CORREL() function in Excel can be used for calculating the correlation between X_1 and X_2 that employs the following well-known mathematical correlation formula

$$CORREL(X_1, X_2) = \frac{\sum_{i=1}^{n}(X_{1i} - \overline{X_1})(X_{2i} - \overline{X_2})}{\sqrt{\left[\sum_{i=1}^{n}(X_{1i} - \overline{X_1})^2\right]\left[\sum_{i=1}^{n}(X_{2i} - \overline{X_2})^2\right]}}$$

where we assumed there are n data points with the following notation used:

- $\overline{X_1}$ and $\overline{X_2}$ are the mean values of the data sets X_1 and X_2 respectively.
- X_{1i} and X_{2i} (where $1 \leq i \leq n$) are the data points of the data sets X_1 and X_2 respectively.

Another statistical method to measure the relationship between two sampled variables by using covariance that has the following mathematical formula

$$COVAR(X_1, X_2) = \frac{\sum_{i=1}^{n}(X_{1i} - \overline{X_1})(X_{2i} - \overline{X_2})}{n-1}$$

Given a data set $X_0 \ldots X_n$, the *autocovariance at lag k* is denoted by g_k and calculated by using the formula $cov(X_i, X_{i+k})$. Hence, g_0 is the variance of X and covariance between consecutive observations in the series starts with g_1. Similarly, the autocorrelation at lag k is determined by using the formula

$$r_k = \frac{g_k}{g_0}$$

The plot of r_k versus k is called *correlogram*. A typical correlogram would look like a bar chart with possible positive, negative, and zero values on the graph.

We have provided the ingredients to determine the variance of the batch averages. The following is a formula known in statistics to calculate the variance of averages of batches:

$$var(\overline{X}) = \frac{var(X_0)}{n} * \left[1 + 2 * \sum_{k}\left(1 - \frac{k}{n}\right) * r_k\right]$$

Noting that now we know how to structure batches, calculate the average of batches, and the square root of variance is the sigma (i.e., standard deviation) term in six-sigma, the six-sigma range can be calculated by using the formula $\overline{X} \pm 3* STDV(\overline{X})$.

6.3 Constraint Optimization Analysis

There are constraints taking place in many problems and it is helpful to write these constraints as mathematical statements in order to understand the nature of the problem description on hand. These constraints might be determined as a part of measurements and the data collected. These constraints are natural occurrences of the problem on hand and help to determine a solution to the problem under specified restrictions of the problem. The way the constraint formation explained above is well known in an area of interest called Operations Research for mathematical formulation. The objective is usually either minimize or maximize and objective function subject to the constraints on hand. In such a constrained optimization problem formulation, a solution to the problem can be attained if there exists a solution. Possible outcomes from such a problem formulation can be no solution, one solution or infinitely many solutions. This theory, optimization theory, is used for solving problems in network theory. The region outlined by the constraints is the domain of the solution and known to be the feasible region. By using optimization problem formulation, problems with hundreds and thousands of variables can be solved. In the case if the problem scale is small, one can use simple programming techniques while more sophisticated programming languages need to be used in much more complicated problems. Some of the well-known software packages that can be used for small scale linear constraint optimization problems are Microsoft Excel and Palisade [26]. Some of the large-scale problems with many data points are MATLAB, Lingo by Lindo, Java, C++, C#, and Python. Noting that there are hundreds of books written in operations research, we don't dive deep into this area of analysis. We recommend the readers to further investigate this area and learn these techniques from these books specialized in this area of interest. In this book, our goal is to just point out that the analysis of the system can be done by using constraint or unconstrained optimization problems with both small and large number of variables. One needs to learn the analysis technique if it is unknown to the user. We do not cover the details of this theory noting that there are books written on this matter and the readers are suggested to find an optimization book that can serve their needs.

6.4 Correlation Analysis

Each variable in a system may relate to another one within the system. For instance, the greater number of computations is needed to determine a problem's solution, the longer that it would take to determine a solution and cause the machine to run longer. By collecting and plotting such data, we can identify the levels of correlation between two or more variables. The types of data correlations include the following:

- Positive correlation
- Negative correlation
- Neutral correlation

The application of correlation analysis can also depend on the CTQ evaluation. An example of a correlation between machine production and the measured CTQ conditions to produce a solution are displayed in Fig. 6.3. Correlation between any two instances is an indicator of how the variables are related to each other based on the data collected and one of the ways to identify this relationship is through straight line fitting (i.e., linear model) to the data set. As demonstrated in Fig. 6.3, there is production data collected from a Machine called X and the associated quality is measured based on the customer's expectations. For instance, CTQ measurement can be based on the expected number of threats detected by the system with an intrusion detection system software installed in

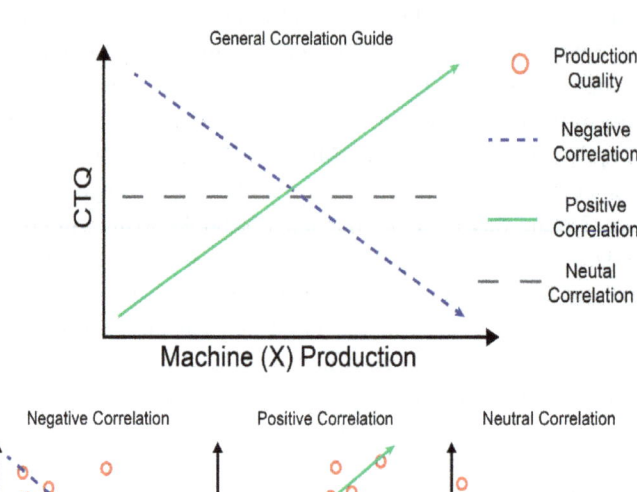

Fig. 6.3 An example for each one of negative, positive, and neutral correlation between the machine X's production for each shift with the output representing the number of defects in millions of items

the machine while the input is the number of threats detected during the corresponding period; Therefore, each red circle in the positive, negative, and neutral correlations in Fig. 6.3 represents such effectiveness of the intrusion detection system by using the associated machine.

6.4.1 Positive Correlation

Given a set of data, positive correlation indicates a positive relationship (correlation) between the input and output variables.

Example (*Incident Time to Close*). The following are the correlations determined for the continuous variables of *Incident Time to Close* data. As can be seen, they are positively correlated.

6.4 Correlation Analysis

correlation analysis	Analyze	Mitigate	Contain	Post-mortem	Full-recovery
Analyze	1	0.98412	0.751973803	0.18633337	0.18633337
Mitigate		1	0.821635906	0.100617947	0.100617947
Contain			1	0.100617947	0.100617947
Post-mortem				1	1
Full-recovery					1

The results indicate a very high correlation among the variables mitigate, analyze, and contain. Analyze and Mitigate have the highest correlation.

Post-mortem also has 100% correlation with Full-recovery variable. On the contrary, both post-mortem and full-recovery cases have very weak correlations with each Analyze, Mitigate, and Contain variables which is about 10–20%.

Using this data, we can conclude that Analyze, Mitigate, and Contain variables are highly and directly correlated with each other as one group while Post-mortem and Full-recovery variables are highly and directly correlated to each other. This is different from the same correlation analysis of the same variables in the medium level where all variables were highly correlated with each other. This indicates a difference in correlation based on these two priority levels.

6.4.2 Negative Correlation

Given a set of data, negative correlation indicates a negative relationship (correlation) between the input and output variables.

Example (*Vulnerability Patching*). Looking at the *Vulnerability Patching* example, the following is observed:

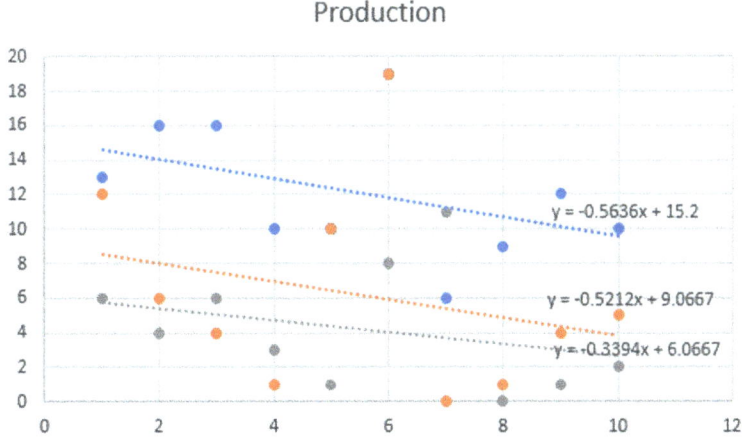

We see a negative correlation between the months (input) and the variables generated for Production Under 30, Production Over 30, and Production Over 45. The rate of decrease in production is similar for both "Production Under 30" and "Production Over 30" data per month, which has a decreasing trend for the most part. "Production Over 45" also has a decreasing trend per month with lower rate than Over 30 and Under 30 data trends. This indicates "Over 45" production to be slower over the period for the most part.

6.4.3 Neutral (no) Correlation

Given a set of data, neutral correlation indicates no relationship between the input and output variables.

Example (*Incident Time to Close*). Looking at the *Incident Time to Close* data, contain (horizontal) versus Post-Mortem in the following image indicates a slope value of – 0.0002 which can be considered as zero therefore these two variables are not related to each other.

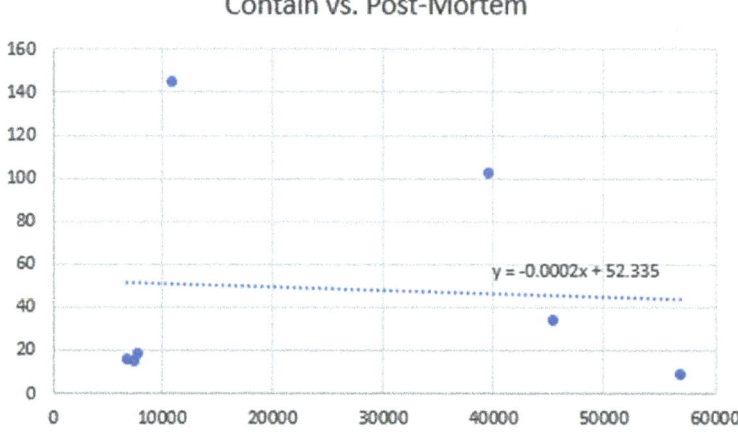

The measured data is ready for your team's analysis right after the Measure phase and it is an important step to decide data categorization for analysis purposes. The data classification/categorization in this section means determining the role of the variables in the collected data on the expected outcome. It is usually essential to decide on the variables' role for the analysis to be implemented. The variable classification means determining the role of the variable within the data set for impacting the output in the way it is desired for the project. For instance, Machine Learning can be used for analysis purposes when too many variables are involved with equally likely involvement of the variables in a problem and the need to determine which variables play the key roles in the solution to the problem with too many variables. Machine Learning requires a trained data set to analyze the

6.4 Correlation Analysis

data. Data categorization can also be done by using Excel with basic statistical analysis as a part of pivot tables, however the role of machine learning and Excel are very different from each other for variable categorization purposes. In some cases, it can be the decision maker's responsibility to tell the six-sigma team depending on their interest in the project after the data is collected. The variable classification of the data can be done by using several methods depending on the nature of the problem:

- **Automated** advanced automated computer-based systems (such as machine learning, search algorithms etc.) that do not require analysis of the data during the analysis work done.
- **Hybrid** methods that include both human decision makers and a basic program used for basic analysis purposes (such as Excel, Minitab, SPSS etc.)
- **Manual** method that does not require any programming during the analysis and the decision maker analyzes the data without any programming work done.

Manual work is rarely seen in analysis of some systems. For instance, a project may require only 5S work requested by the customer and the team might have to use qualitative judgement to ensure that the improvements are made accordingly. In such a case, human factors concepts and lean methodologies may appear to be much more important compared to statistical measurements for data collection. It is possible to use automated and hybrid methods even if a project might require only human factors and lean systems, however the team may not have background in the way to quantify such data.

For cybersecurity applications, both the automated and hybrid methods can be easily seen in six-sigma projects due to the amount of data collected and the decision maker's goal. Machine Learning can be a method of practical use for data sets with many variables. For instance, if there are 30 variables with 150 numeric instances (you can think of it as 30 columns and 150 rows) then the data set is large to apply a hybrid or manual method for variable analysis. The hybrid analysis can be done by coupling variables and observing their relationship with each other or by using pivot table in Excel; this method can help to determine the impact of the variable on the output. Other practical software packages such as Minitab and SPSS can also be used. More advanced software packages that require programming experiences such as R, C++, Java, Python and MATLAB can be also used, however they require good working knowledge of the script and algorithmic development before the analysis can be started. In some cases, the six-sigma team may lack machine learning experience, and the team might have to hire a consultant to determine a machine learning solution.

6.4.4 Strength of Correlation

Strength of correlation as an indicator of how two random variables X_1 and X_2 relate to each other is very helpful and such correlation's categorization can be evaluated by grouping strength levels of correlation within the range between -1 and 1 as it is shown in Fig. 6.4. A meaningful way to classify correlation is by introducing levels of strength based on numerical levels of correlation strength as follows:

- **No Correlation/Relationship**. In the case when the correlation value is either zero or almost zero, the variables have no correlative relationship. The meaning of "almost zero" depends on the data used therefore it is up to the reader to identify the definition based on the project's scope. For instance, a data set may require the definition of less than 0.001 to be almost zero while another one may require it to be 0.000000001 depending on the data set used.
- **Positive and Weak Correlation**. Positive weak correlation's range assignment also depends on the data set and the project. For demonstration purposes, it is assumed that two variables have positive weak correlation when the correlation value is anywhere between 0.001 and 0.5 below.
- **Positive and Moderate Correlation**. The range of values assigned to positive moderate correlation also depends on the project. For demonstration purposes, it is assumed that two variables have positive moderate correlation when the correlation value is anywhere between 0.5 and 0.75.
- **Positive and Strong Correlation**. The range of values used for positive strong correlation also depends on the data set and the project. For demonstration purposes, it is

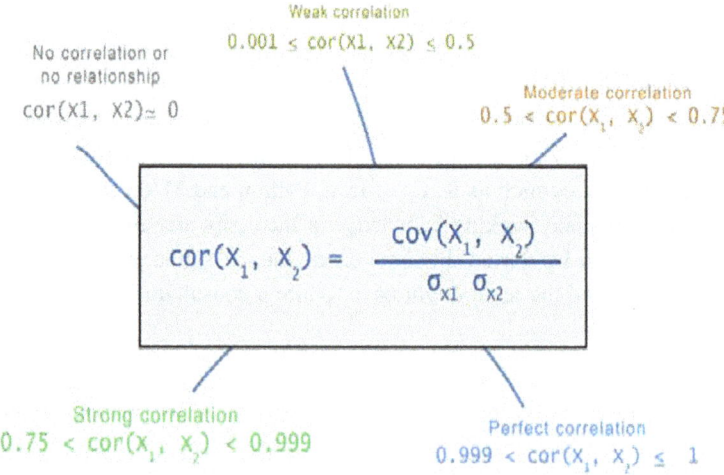

Fig. 6.4 Correlation formula outlining the strength categories of correlation

assumed that two variables have positive strong correlation when the correlation value is anywhere between 0.75 and 0.999.
- **Positive and Perfect Correlation**. The range of values assigned to positive and perfect correlation depends on the data set and the project. For demonstration purposes, it is assumed that two variables have a positive and perfect correlation when the correlation value is anywhere between 0.999 and 1.

The positive correlation groupings outlined in Fig. 6.4 work for the negative correlation classification of the variables.

6.5 Design of Experiment (DoE) Analysis

Design of experiments (DOE) is defined as to plan, conduct, analyze, and interpret controlled tests to evaluate the factors that control one or more values of a parameter or group of parameters. The analysis of strategy used for experimented measurements that are planned and executed may provide a great deal of information about the effect on a response variable due to one or more factors. Initially decided KPI may be realized to need modification as the data is collected. One important aspect of Six Sigma project is the need for different mindsets to gather and answer questions from different angles. The design of DoE needs to be structured carefully, and the inputs that impact the output (which is the response of the system) need to be determined with the scope of the six-sigma project.

Example *(Incident Time to Close)*. The data of *Incident Time to Close* originally contained 12 data columns. The priority levels Low, Medium, and are used for separating the original data set into three different subsets for further analysis of each individual priority level. As analysis of the data is further developed, one weakness of the collected data, as a result of DoE analysis, is the missing information on the impact factor (say on a scale of 5) of not handling the vulnerability on time. If this information was incorporated into the data set, then the analysis of the data can be further improved by first reacting to the highest impacting vulnerabilities for improvement as a part of six-sigma. In addition, the dollar cost of the impact can also be collected, if possible, to impact the dollar cost of the issue faced.

6.6 End Goal Analysis

The "end goal" in this section means the defined problem's expected outcomes in the way that the problem is defined by the six-sigma team in Define phase. Success in an end goal means fulfillment of the minimum threshold for expected success in the way the problem's solution is formulated, therefore the end goal excludes delighters.

Subject to the interest of the six-sigma team, an end goal analysis can be accomplished in the Analyze phase by checking on the following:

- Check the feasibility of reaching the expected end goal. The team can try to predict whether the expected outcomes for the project can be achieved or not because of the analysis accomplished.
- In the case if the analysis appears to be meeting the expected end goals of the project, you will be good to move on to the Improve phase once the analysis is completed. If the end goal analysis indicates that an expected end goal cannot be reached, it might be essential to revisit either the Define or Measure phase:
  - In the case if you need to revisit the Define phase, you may have to redefine the problem and recollect the data (if needed) depending on the level of changes on the problem's statement.
  - In the case if you need to revisit the measure phase, you may need to recollect or collect more data.
- Once the six-sigma team ensures that the Define and Measure phases are restructured (if needed) then the team is ready to move on to the Improve phase.

The end goal analysis can have a variety of purposes; however, the main goal is to check the comprehensiveness of the first three phases of DMAIC and their relation to the end goal in several different ways.

As an example, suppose a grocery store wants to understand the nature of the sales and wants to improve the sales by marketing the store during certain periods of time. A six-sigma team is employed by the owner of the grocery store and asked to identify periods of time during which it makes sense to advertise the store and increase sales. Suppose the six-sigma team defines the problem without any time frame considerations of sales and starts collecting data randomly and analyzes the data without time frame consideration. The problem with defining the problem in such a way can be the periodic nature of grocery store sales. After collecting and analyzing the data for six months, the six-sigma team may realize the fact that there are differences between weekdays and weekend sales. The six-sigma team may need to redefine the problem in a way to include a timeframe and periodic nature of sales by determining periods of similar sales. In addition, it will make sense to collect a minimum of year-long data to have sense making results for the end goal of the project (because special days such as Thanksgiving, Labor Day etc. happen once a year and your data may not cover them). In the case when data already exists, the

6.6 End Goal Analysis

way to proceed with analysis can be much easier if the data is clean enough to analyze (which is usually not the case therefore you may need to clean the data) and collected in a way to meet the end goal(s) of the team's project. The team can continue analyzing the data as collected from the beginning and not waste time with the project's end goals to determine the periodicity of the sales. Monthly and weekly sales' (with differences between weekdays and weekends) analysis might be critical to understanding the nature of the sales in this project, which is not included at the beginning of the project. The team may also realize during the analysis phase that there is a need for surveying customers to observe what type of marketing would interest them to shop at the grocery store and when they shop at the grocery store. In such a survey, questions on the periodicity and days that the customers' shop also needs to be investigated.

Some of the missing concepts that could impact the end-goal analysis of a six-sigma project are the following:

- Periodicity
- Broad coverage of concepts for the project (i.e., is the project broad enough to cover areas of interest in the workplace, materials, workers etc.)
- Right sample size for different subgroups (i.e., was the sample size right for all subgroups that is covered in the project?)
- Changes in the system that relates to the project (i.e., are there changes in the operations of the project since the beginning of the project?)
- Is the end goal realistic? (i.e., The Analyze phase helps to oversee whether the end-goal of the project is feasible or not. If it is not, then it might be essential to revisit the Define and Measure phases)
- Are the customer expectations and analysis of the collected data appearing to result in expected end-goals statistically? For instance, the current analysis of the system may indicate 65% success in fulfilling the customer needs and the system's maximum capability (i.e., maximum utilization of the system) may allow up to 75% success in the way that the customer expects. The problem might be initially defined as having 95% success for the project. Unless the customer agrees to change the system components to yield 95% success (if it is feasible to do so as the system cannot jump to 95% accuracy) the problem needs to be redefined and remeasured. In this way, the customer realizes that the initial high expectations are infeasible in the system. Refinement of the problem may also change the measured instances for the problem. (i.e., would it be possible to derive statistically significant results that meet customer's expectations from the project and goals?)

Example 1 (*Security Tool Coverage*). We mentioned previously that the customer expectation in the example *Security Tools' Coverage* is around 85%, which turned out to be realistic based on the analysis results to be explained later. For instance, in April, NIPS, TVM, EDR, and SIEM cover an additional amount of 87% of the missing tools. If the company has the financial means, then this need can be covered. This data set had 13,944 data points which is sufficiently large for a data set.

Example 2 (*Incident Time to Close*). Considering the *Incident Time to Close* example, there are 10 high priority data points, 34 medium priority data points, and 348 low priority data points collected. The need for more data points on high and medium priority levels is clear in the way the data is collected. This may require the period to be longer, however it is worth noting that the results will be stronger with more data included for all priority levels.

6.7 Equal Variance Testing

Variance is an important statistical concept for determining the distribution of a data set. I.e., variance helps to determine how much the data statistically varies from the average (i.e., mean) value of the data set. Measuring variance is particularly important to test if the distribution of the data is away from the mean or not, and as you can recall, this is the focus of six-sigma noting that the sigma is the standard deviation value (which is the square root of variance). In the case when the variance is same for data sets that have same or similar nature, it is a positive sign for the system's analysis. For instance, after determining a distribution to cyber threads, variation from the average number of attacks during a specific year by using five different defense systems can be simulated by using these different cyber defense mechanisms. Assuming approximate normal distributions of the collected data, the measurements can be analyzed by using average and variance/ standard deviation values. The differences between the identified values would indicate the value of the method used. In the case when the average values are same but the standard deviation different, the significance can be determined by using the standard deviation value differences that simply is what we called here "equal variance testing".

As can be seen, equal variance is beneficial in many ways for many different purposes. In the case if the variance is high for the same data set mentioned above, it is an indicator of issues with the defense mechanism noting that responses to threads is not under control by using the cyber defense mechanism. In the case when a system is not under control, it simply means we cannot plan the corresponding thread responses properly, which also means the system's response is not under control. The best outcome from a variance test would be a low variance value with possibly equal variances for similar data sets.

6.8 Failure Mode and Effect Analysis (FMEA)

It is important to decide the level of priority to deal with the areas that need to be improved. During the analysis of such areas, the general information that relates to the project needs to be written down to determine the mode in which the failure occurs and the analysis of the failure's effect on the overall system needs to be determined. Such general information is project dependent however there are certain process components that can be stated in general terms that take place in FMEA:

- State the location, process step, and function to clarify what or where we are dealing within the system.
- State the expected standard for failure to occur.
- Analyze the impact of the failure on the output.
- Determine the severity impact of each possible failure on the overall relevant system.
- Calculate or determine the failure frequency based on the collected data by using the measured data and measurement standards employed for the system.
- Design a cause-effect diagram for each type of failure of interest to determine what can be used for eliminating such a failure in place.
- The risk priority matrix components such as severity, frequency and failure detection rates can be determined as a part of the risk prioritization analysis.
- Determine the actions by reducing the risks that cause failures with their costs and benefits.
- Identify the dates of actions taken to eliminate the corresponding risks and apply 5W+H to determine details on the corrective actions taken.
- Investigate to find out how much the customer cares about the reduction of the risk and how the customer reacted to the reduced risk with the corresponding method in the past (if applicable).
- The impact of the corrective actions in overall system's performance.
- After corrective actions taken in the improvement phase, apply Measure to recollect data and redo the analysis on the new system from the beginning with the update on the risks and failure possibilities, rates, actions taken for reducing failure and overall customer satisfaction.

You can have a custom-made rating at a level of 5- or 10-scale so that the severity level can be designed depending on the scope of the project and measurements. This rating can also be consulted by the customer to determine the level of satisfaction. For instance, the scale rating can be determined by the customer as follows:

- 1-scale: The failure in the system is observed to be somewhere between 10 and 20%
- 2-scale: The failure in the system is observed to be somewhere between 8 and 10%
- 3-scale: The failure in the system is observed to be somewhere between 3 and 8%

- 4-scale: The failure in the system is observed to be somewhere between 1 and 3%
- 5-scale: The failure in the system is observed to be somewhere between 0.001 and 1%.

As can be seen in this example, the scaling is based on unevenly distributed percentages of failure, and it allows the six-sigma team members to follow the evenly distributed scale. This evenly distributed scaling also helps with construction of figures and eases the analysis.

In the case when the customer leaves the scaling choice to the six-sigma team then the six-sigma team needs to first investigate the best practices employed by others in the past and search for standards or potential standards that might have to be applied as a part of such a system. For instance, ISO 9001 can be used as a standard, however the scaling might have to depend on the six-sigma team and data collected. One other method of handing such scaling can be statistical distribution and the corresponding standard deviation depending. Recall that the meaning of Six Sigma is a certain percentage of accuracy. One can apply the 1-sigma, 2-sigma, ..., 6-sigma standards to scale the data. In the case when we have batches of data, one can consider using the batch means method. In all the cases explained above, it is important to incorporate what the customer considers as success and what relevant standards can be considered to eliminate waste and risk.

The scaling for severity, frequency and probabilities for risk detection need to be defined separately and need to be made sense from the customer's and process' perspective.

6.9 Interaction Table

Observation of interactions between system components is one of the crucial and value adding Six Sigma applications on data sets. The interaction table can be used to understand the interaction between the system components of the process or service in place.

Suppose we observe a system's data with three machines, M1-M3, with the workflow moving first from M_1 to M_2 and then from M_2 to M_3: Hence, in this system, output of M_1 moves to M_2 as the input for further processing and the output produced by M_2 becomes the input of M_3 for further processing. Each machine is utilized for detecting the same threats in a consecutive order. This data is a simulation of a system for understanding the effectiveness of three different intrusion detection systems (IDS) provided by three different vendors placed in three different machines for configuring the effectiveness of threat detection and ability to response to such threats on the same data. Observations in the table represent the number of issues determined during the four consecutive observations with each observation's completion during consecutive months. It is known that there were updates each month throughout the four observations/months. The last column displays the total number of threats identified in the network within that month. For

6.9 Interaction Table

instance, during the first month of observations, M_1 was able to detect 188 of the 200 threats while M_2 detected 187 of the 200 threats and M_3 detected 185 of the 200 threats.

	M_1	M_2	M_3	Total # of items
1st observation	188	187	185	200
2nd observation	240	239	236	250
3rd observation	199	208	205	210
4th observation	288	299	299	300

Given that the processing relied on the improvements of the software during each observation, our focus needs to be on the following interaction with the system:

$1st\ Observation \rightarrow 2nd\ Observation \rightarrow 3rd\ Observation \rightarrow 4th\ Observation$

After analysis of the data, the interaction table can be used to understand the interaction between the mechanisms from an IDS's ability to identify threats perspective by using the following signs:

%   The increase in the number of threats from one observation to the other.
&   The decrease in the number of threats from one observation to the other.
—   No change in the number of threats from one observation to the other.

We can organize the columns of the interaction table to represent the positive or negative relationship between the changing capabilities of the IDS. We can reorganize this table as follows:

	1st observation (%)	2nd observation (%)	3rd observation (%)	4th observation (%)
M_1	94	96	99.5	96
M_2	93.5	95.6	99.05	99.67
M_3	92.5	94.4	97.62	99.67

The corresponding interaction would be like the following:

	1st–2nd observation	2nd–3rd observation	3rd–4th observation
M_1	%	%	&
M_2	%	%	%
M_3	%	%	%

For instance, in this table, during the first three observations, the first software in M_1 had increasing threat response rates that decreased from 3rd to 4th observation. We can extract several causes to this issue by answering the following questions:

- What caused the significance of the changes for the software tested in machine 1 to be less successful later during the observations?
- Which of the software had the highest success and why?
- Where can we apply improvements in this system to make the threat responses better?
- How can we improve the system to design a better IDS response system?
- Why is M_2 more successful than the other two IDS?

Answering these questions would provide a very useful insight.

Example (*Vulnerability Patching*). Consider the following interaction table for the *Vulnerability Patching* example in Fig. 6.5.

Color coded observations in the table represent the number of changes from one state to the other. For instance, there is a decrease in the number of cases in row 2020-01 from Production Under 30 to Production Over 30 with an arrow pointing down in a green box. Similarly, there is an increase in the number of cases on row 2020-08 from Development Over 30 to Development Over 45 with an arrow pointing up in a red box. There are three yellow boxes with no change in three cells that indicate no change from one state to the other.

If we take a comprehensive look in the data set for the ten-month period, we can see that the least increase in instances occur from Production Under 30 to Production Over 30 which is a good indicator that the Production decayed over the 10-month period

	Production Under 30	Production Over 30	Production Over 45	Development Under 30	Development Over 30	Development Over 45
2020-01		↘	↘		↘	↗
2020-02		↘	↘		↘	↘
2020-03		↘	↗		↘	↘
2020-04		↘	↗		↗	↗
2020-05		→	↘		↘	↗
2020-06		→	↘		→	↘
2020-07		↘	↗		↗	↗
2020-08		↘	↘		↘	↗
2020-09		↘	↘		↘	↘
2020-10		↘	↘		↘	↘

Fig. 6.5 An example of an interaction table

which can be a good indicator that the issue is handled in an effective way (but further analysis is needed by applying other analysis techniques!). Looking at the transition from Development Over 30 to Development Over 45, there is dramatic decay. This could mean that the Development is not needed as much, or the issues are lowered therefore there is not as much development needed. In this case, furthermore analysis is needed by applying other Analyze methods.

6.10 Interface Analysis

Interface analysis is the analysis of transition from one functionality to another. During interfaces there can be waste occurring easily depending on the conditions. For instance, in cybersecurity applications, different parts of the network may have different ways to handle cyber-attacks that may result in different processing times.

In cybersecurity applications, different parts of the network may have different ways to handle cyber-attacks that may result in different processing times. In the case when transitioning from one network security handler to another, the time calculated as a result of the interface analysis can be much larger than expected which could result in time waste. If we consider the *Vulnerability Patching* example, the value 19 highlighted in pink indicates an issue in transition from Production Under 30 to Production Over 30. Highlighted yellow areas indicate issues in the interface while transitioning from the prior state to the current state.

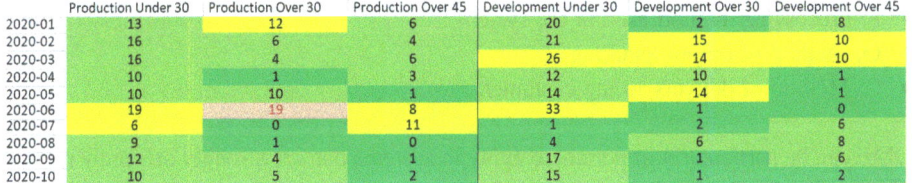

6.11 Maximum System Capability Analysis

In a typical system, it is natural to wonder about the maximum possible work that can be completed without any waste occurring in a system. We usually need to know this to determine the top level (i.e. maximum) of a system's capability that we should not exceed. For instance, considering the *Security Tools' Coverage* example, the entire data sets' tool coverage need is the maximum numbers for each one of the 8 tools considered. Depending on the interest of the period, either each month or the three-month period can be considered as the maximum tool and inventory coverage needs. For instance, in the case of monthly considerations, the following is observed for the *Security Tools' Coverage*:

- Considering April, the most needed tools in the inventory (that are missing) are NIPS, TVM, EDR, and SIEM, which account for about 87% of the needs. This can be useful for initial action purposes.
- In May, CMDB needs to increase compared to April. Compared to April, NIPS, SIEM and TVM are still major needs. CMBD needs are added to this list while EDR needs are less effective compared to April. The total instances needing CMDB, EDR, MEVS, NIPS, and TVM account for about 94% of May needs.
- In June, CMDB needs increases by three times compared to May. All except CDP and NIPS are major needs that cover about 89% of the needs.

The maximum capacities are calculated based on the inventory need for each tool.

6.12 Overall Methodology Effectiveness (OME)

Overall Methodology Effectiveness is a way to measure the method used and helps to evaluate the methodology that related to the defined problem in the Define phase. Recall that IPO or SIPOC can be the basic building block of product or service, and the "method" can be anywhere in IPO or SIPOC. For instance, some of the methods that can be considered include

- S: The method of purchasing from the supplier
- I: Effectiveness of the method used for achieving the input
- P: Processing method used in the system
- O: Method of output handling
- C: Method of delivery to the customer.

OME can be particularly important to apply in systems that are strongly driven by methodologies. In such systems (or this could be the defined problem while the system can be much more sophisticated than the problem), the involvement of equipment, hardware and manpower can be easily ignored due to their irrelevance to the defined problem. The solution to the six-sigma problem at hand might be driven by the methodologies and hardly anything else. In such a system, the methodology can be evaluated after quantifying the tasks completed. For instance, considering the response time of a system to cyberattacks, the method of response process can be evaluated by the time it takes for the software to respond to the attack by quantifying it while the method of output handling can be evaluated based on a numerical scale 1–5 that ranges from weak to strong. OME can be calculated by using the formula.

$$OME = ((good\ output\ received\ using\ the\ method)/total\ output) * 100\%$$

6.13 Pareto Chart Analysis

where total output is speed of the method multiplied by the time used for the method and the corresponding output amount.

Example Suppose there are 50 breaches as a result of 1000 cyberattacks happening in 8 h and each attack is handles in 30 s on average as a result of using method 1 while there are 65 breaches in 1500 cyberattacks happened in 10 h with each attack handled in 40 s. OME1, corresponding to method 1 is

$$OME_1 = \frac{1000 - 50}{480 * 0.5} = 3.958$$

while the corresponding second method OME2 is

$$OME_2 = \frac{1500 - 65}{600 * \frac{2}{3}} = 3.588$$

Therefore, we can conclude that the first method is much more effective in handling cyberattacks based on its credentials.

6.13 Pareto Chart Analysis

It is possible to combine a bar graph and the accumulative behavior of data by using a Pareto chart which is organized as a bar graph with accumulation of points in the bar chart is designed as a curve from left to right of the chart. A Pareto chart rearranges the input in a way to categorize the data into ranges of values that are represented by bars while the percentages of the accumulated data are displayed on the left side (also called the third axis) of the graph in an Excel spreadsheet while the output values are displayed on the left side of the Pareto chart. A Pareto chart is useful for observing the increasing trend of the data values based on the subgroup displayed by the bars.

Example (*Security Tool Coverage*). The following two Pareto charts and their distributions are designed for the Security *Tools' Coverage* data. The following Pareto chart is designed for the tool-target match; Given tool is available; the target also contains the tool that is needed.

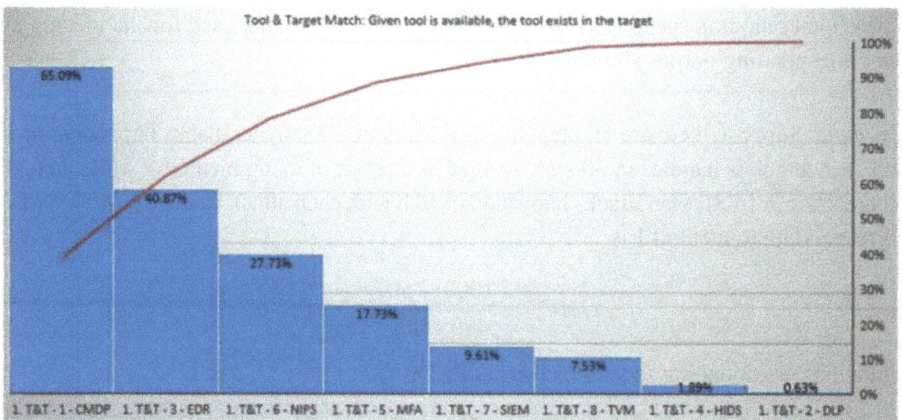

The following exponential distribution is structured for the tool-target match.

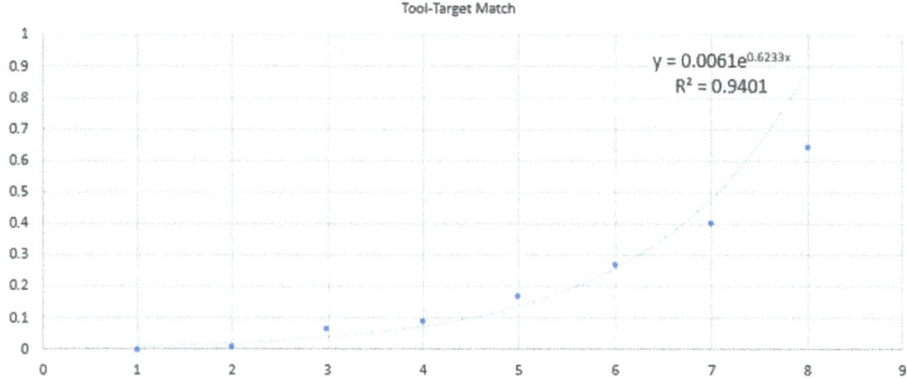

$y = 0.0061e^{0.6233x}$
$R^2 = 0.9401$

The following Pareto chart is designed for the targeted tool needed in the inventory while the tool does not exist for the corresponding operating system. This Pareto chart indicates the major need for NIPS in the mismatch while CMDP and TVM follows as the secondary and tertiary needs in this list. As it can be seen, there is no need for DLP and the red curve that representing the cumulative percentage becomes a horizontal straight line as it reaches to DLP tool.

6.14 Process Efficiency Analysis

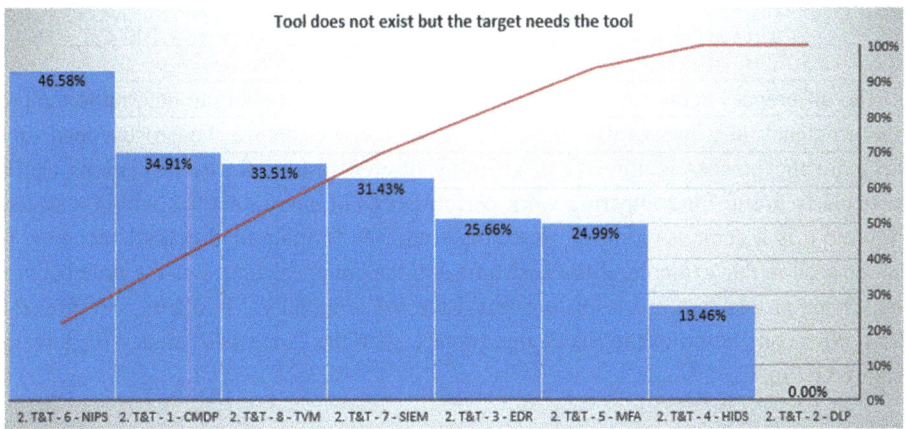

This need is quantified by the following linear regression to observe the rate of increase from one tool to the other. The input axis in the below regression line is designed to have an increasing trend. For instance, DLP tool is represented by 1 in the input axis below while HIDS tool is represented by 2; This order of the inputs in the below Figure are the inputs of the corresponding Pareto chart above from left to right.

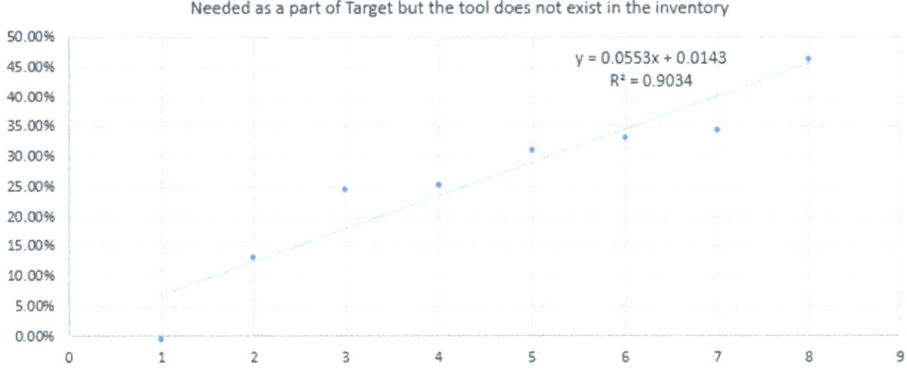

6.14 Process Efficiency Analysis

Process improvement depends on NVA reduction and increase in the VA. A system that has such changes would be able to process much more effectively with an increase in the system's capabilities. Analysis of a process not only depends on the determination of the VA time to the process but also meaningful when it is compared with the lead time of the process. Therefore, efficiency of a process can be calculated by using the following formula:

$$Process\ Efficiency(\%) = \frac{Value\ Added\ time\ of\ the\ process}{Process\ Lead\ Time} * 100$$

The differences between lead and value-added times would help to determine the process efficiency in a measurable way. The closer the VA time to the process lead time, the more the process is efficient in attaining desired results. Such calculations can be particularly useful for comparing work performances of similar entities within a system. Figure 6.6 is a demonstration of an example and the formula of Process Efficiency. As can be seen in this example, if the lead time is three hours calculated as the time between the sorting process and truck loading and if the VA is initially 2 h and then improved to 2.5 h after completion of the Six Sigma project, then the Process Efficiency is increased from 66.67 to 83.33%.

Example (*Vulnerability Patching*). Considering the Vulnerability Patching example, Productions under 30 and 45 as well as Developments Under 30 and 45 can be considered as the value-added times while both Production and Development over 45 might be considered as NVA. Assuming that "the Production Over 45" is ended on average around 65 and Development Over 45 is closed around 75, the corresponding PEs can be calculated as follows:

$$PE_{production} = \frac{45}{65} * 100\% = 69.23\%$$

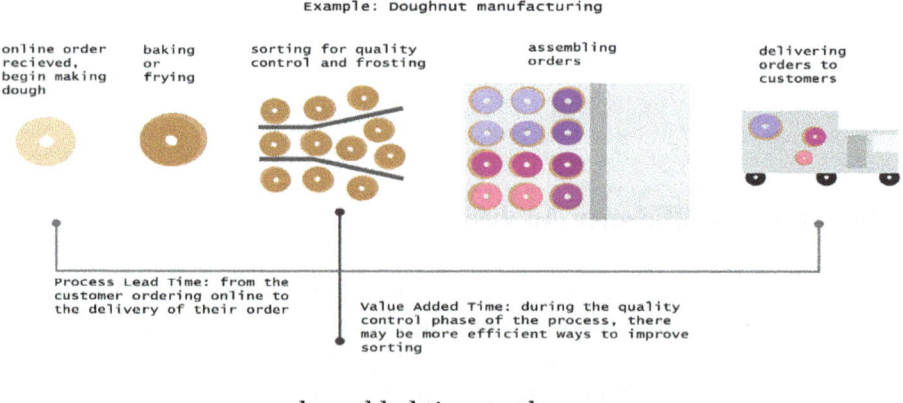

Fig. 6.6 An example of PE analysis and its formulation

$$PE_{development} = \frac{45}{75} * 100\% = 60\%$$

This analysis can guide the six-sigma team to determine the maximum needed maximum average production and development for a targeted percentage of process efficiency that can be stated by the customer.

6.15 Right Lean Techniques' Selection

Selection of the right Lean technique for improvement and organization of activities (including tasks to be accomplished by humans, machines, items, system components, etc.) to maximize value in the process by reducing waste is needed. Reducing all the existing waste altogether is likely to be impossible in many systems hence application of lean techniques continuously may be needed. As a simple , regular updates and maintenance of the equipment such as cleaning of computers require scheduled maintenance and a proper procedure to follow with certain steps. In addition, it is essential to look for application of the lean techniques to identify additional waste; therefore, lean techniques usually require a periodic approach to enhance improvement. In this regard, a cyclic work approach is needed with the cycles of the work to be completed following proper procedures and techniques. Hence, the Six Sigma project's outcome would require outlining this continuous improvement application. Six Sigma with the incorporation of Lean techniques is called **Lean Six Sigma**. Usage of Lean techniques is particularly helpful for incorporating tested improvement techniques after understanding their strengths and using them appropriately in Six Sigma projects. There are well developed and known Lean strategies introduced by many researchers and some of these will be briefly explained in this chapter. Kaizen event and Value Stream Mapping of business operations that we covered previously are also Lean techniques used in improvement efforts. The following are the Lean techniques that we will cover in this section:

- Heijunka
- Jidoka
- Just in Time (JiT)
- Kaizen
- Poka-Yoke
- Value Stream Mapping (VSM).

Even though we cover the Lean methodology as a part of the Improve phase, its' usage is not limited to this portion of DMAIC. Some of the Lean methods such as Value Stream Mapping and JiT may require time studies and initiate from the beginning of the project with the Define phase and continue all the way throughout the project up until the end of the Control phase. The VSM is introduced as a map to visualize the operations with a certain level of detail required by the project. The Measure phase incorporates VSM related measurements needed while Analyze phase requires the analysis of the measurements attained for the VSM to be able to identify improvement opportunities with the analysis of the measured process and waste data (such as mean, SD, variance, takt time etc.). The Improve phase requires an update to the VSM to reflect the changes outlined for the system after applying the improvements within the system. Hence, the VSM to be covered in the Improve stage is a map that reflects the improvements within the system after eliminating NVA. The waste can be marked on the VSM to be able to apply the waste reduction in the Improve stage. In some cases, it may need to be used as the project progresses as a sub-project. For instance, application of a Kaizen event may be needed during a Six Sigma project to advance to the next level of analysis.

6.15.1 Heijunka

Heijunka, or production leveling, is a method to even out production by observing its uneven nature. This method requires maximizing the uneven production and minimizing the variation throughout the process. The uneven nature of the process can be simply observed through a bar chart. Considering the Vulnerability Patching example, Heijunka can be applied to Production Over 45 which can be considered as NVA with both Production Under 30 and Production Over 30 are adjusted for Production Over 45 to be leveled off. An equal workload per month can be implemented as much as possible.

6.15.2 Jidoka

Jidoka, or autonomation, is a Lean method typically used in manufacturing. It is a way to quickly react to out-of-control cases by halting the operation. Even though Jidoka may not seem to be appropriate for applications in six-sigma application of cybersecurity, it can be applied. For instance, the result of a six-sigma project may indicate an unleveled workload that may cause issues in the thread handling. A six-sigma team indicating that the thread handling is below the expected accuracy of the customer would result in Jidoka; the customer could immediately act and change the method of handling threads in the Vulnerability Patching example. This also applies to the *Security Tools' Coverage* example. Once the tool-inventory is determined to be mismatched 100%, Jidoka can be applied and a structured method of resolving this issue can be initiated.

6.15.3 Just in Time (JiT)

Just-in-Time is a management strategy that designs a system in a way to utilize raw-material orders directly from supplier to production schedules. The goal of JiT is to manage inventory to increase efficiency and decrease waste by receiving goods only as they need them for the production process that may reduce inventory costs. Forecasting demand accurately is important for the producer.

An example of JiT in cybersecurity applications can be seen on the *Security Tools' Coverage* example. Noting that the inventory numbers are 3,643, 4,094, and 6,207 for April, May, and June respectively, one may predict the following quadratic model in the figure below to forecast the inventory tool need in July by calculating

$$y(4) = 831 * 4^2 - 2042 * 4 + 4854 = 9,982$$

One weakness of JiT is its prediction dependent nature. If this prediction is lower or higher than the expected number, then the tool-inventory mismatch will continue to occur even after Six Sigma project. A control mechanism by using control charts would be needed with an appropriate JiT methodology needed for such an environment.

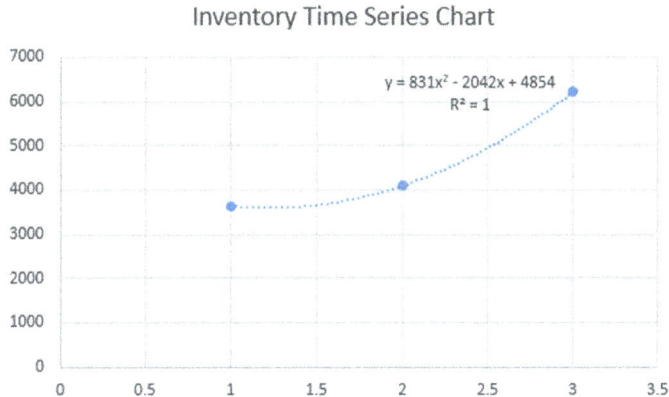

6.15.4 Poka-Yoke

Poka-yoke is a Japanese term used for mistake-proofing. The goal is to design a system in which the mistakes are human errors avoided to eliminate waste by preventing, correcting, or drawing attention to these errors as they occur. Recall that we determined 100% tool-inventory mismatch in the *Security Tools' Coverage* example. The six-sigma methodology used to determine this inefficiency of mismatch is due to the analysis techniques not used and six-sigma not being applied. The Poka-Yoke application in this setting can prevent and correct the mismatch over the following months. Causes and effects need to be outlined as needed when determining the mismatches.

6.16 Resource Analysis

We define resources to an entity in this section as the direct feeding mechanisms to the entity. For instance, if there are jeans produced in a factory, the zipper to be placed on the jeans directly arrives from the corresponding manufacturer and machines accordingly. The entity in this case is the machine that saws the zipper while the resource is the manufacturer. In the same production, a machine can be the resource to another machine as a part of the flow of production. Therefore, the resources are entity dependent. The success of a system can be analyzed by the progressing parts of the product. If we consider an item produced by using 10 machines and the item is produced by passing the product from one machine to another. Suppose there is only one raw material that goes into the first machine and then goes all the way up to machine 10. Each machine applies a different process to finalize the product and the piece of raw material that goes in comes out of machine 10 as a single unit as well. In such a case, if each machine has 95% successful output (meaning there is 5% defect) for the machine they apply, it is likely that the success rate will be $(0.95)^{10}$ or less. It is amazing to see that the defect rate could be extremely high in such a case.

In the case of cybersecurity applications, the network has many connections and failure in the network could result in inefficiency of the system. Noting that a package sent in the network goes through many routers and these routers have levels of success in handling such cases, the reliability of the routers should be very high to be able to handle package delivery. In addition, the reliability of the network depends on the success of delivery of the package to the cloud. The analysis of a network could be done by using resource analysis. The output of an entity is a success indicator of the entity for the next step to be taken in the process.

6.17 Root-Cause and Cause-Effect Analysis

We need to first recall that the end goal of a six-sigma project is the elimination of a determined waste to improve a system. The analysis of the measured data along with the time study helps us to understand the system better and determine the waste that arises. Assuming that there are certain subcategories that appear as the "roots" of the waste, these subcategories contain the "causes" of the problem we are facing. Root-cause analysis is also known as cause-effect analysis or fish-bone diagram. For instance, in cybersecurity, if there is a cyber-tread handling weakness then the possible roots of this problem can be related to

6.17 Root-Cause and Cause-Effect Analysis

- People/Staff
- Machines
- Methods
- Materials
- Nature
- Measures.

Related to all these roots, there are potential causes to occur. For instance, examples of the root "Nature" can contain the following causes:

- Connections between machines (server and local machines).
- Malicious activity levels (occur naturally).
- Volume of staff communication.

If you are not familiar with cybersecurity operations, then you may have a hard time following several of the effects listed above such as malicious activity levels that occur naturally. In the case when too much malicious activity exists, additional threat handling mechanisms may have to kick in to strength the defense, and additional security measures can cause challenges to employees that may cause time waste for completing tasks.

In cybersecurity applications, suppose there are too many attacks in a banking system, and this causes too much money to be wasted by the bank. Possible roots that can result in this waste include the following categories:

- Users
- Software
- Cloud
- Network
- IT.

These roots can contain many "causes" that can result in the waste; the observations in the system during the measurements are expected to help you outline the specific causes to the waste. For instance, as a result of measurements, you may suspect that the age category 45–65 has the maximum number of breaches to the bank's system as a result of the analysis of the age categories you focus on your statistical analysis of the cyberattacks. There can be other specific roots that you can determine along with the specific causes that arise. Only the roots and causes that relate to waste should be recorded in the root causes analysis, otherwise we overload the roots and causes with irrelevant data, and this causes waste within the observations we have and the corresponding analysis. On the other hand, we need to write down the roots and causes that are suspected to cause the waste even if we don't know whether they can cause the waste or not; This is because you are part of a team and what you suspect may be known by another team member who can provide you more information on your suspicion.

The way to determine the roots and causes of a waste is to walk through the entire pathway that can cause the waste. This is just like daydreaming in which you mentally picture all the steps that take place in the system, and you walk through the steps in your mind with the details of all the problems that can arise and the details of the considerations for success. This is one of the reasons that time study is very important to understand the details of system in place. As you are a part of the system and experience all the steps of the system, you should have a good working knowledge of what can happen and how it can happen with success and failure specifications. It is then important to take notes and fill out the root-cause chart to not miss anything about the problems faced.

Example (*Incident Time to Close*). Returning to the *Incident Time to Close* example, the following are observed to be the causes for medium-priority level incidents to occur:

Inadequate risk analysis	79.41%
Unplanned project	2.94%
Security configuration error	2.94%
Threat actor targeted individuals	2.94%
Unexpected internet activity	5.88%
No impact	2.94%
Vulnerability exploit	2.94%

with the corresponding follow-up actions outlined:

Review	79.41%
Closed	4.08%
Identify	4.08%
Review	6.12%

In the case of low priority, the following are observed to be the causes for incidents to occur:

Inadequate risk analysis	24.75%
Unplanned project	1.49%
Threat actor targeted individuals	1.49%
Unexpected internet activity	6.93%

(continued)

(continued)

No impact	63.61%
Vulnerability exploit	0.25%
Threat actor targeted individuals	1.49%

with the corresponding state analysis

Closed	1.26%
Identify	4.77%
Review	88.44%
Analysis	3.27%
Contain	2.26%

In the case of low priority, the following are observed to be the causes for incidents to occur:

Inadequate risk analysis	80.00%
Unexpected internet activity	10.00%
Security configuration error	10.00%

with the corresponding state analysis

Noi	20.00%
Unidentified	10.00%
Other	40.00%
Malicious code activity	10.00%
Phishing	20.00%

6.18 SWOT (Strength-Weakness-Opportunities-Thread) Analysis

A system in place would naturally have strengths, weaknesses, opportunities (to further benefit from the system) and threads that can be analyzed to implement SWOT analysis. If you choose to use SWOT analysis in your six-sigma project, then strengths and weaknesses indicate the two edges of the system's analysis components while opportunities and threads represent the associated impacting factors in the system. SWOT analysis can be viewed in a coordinate system structure after relevant numerical analysis is completed. There are basic steps that can be followed to implement SWOT analysis:

1. Identify SWOT components in the system that are relevant to the defined problem. Categorize SWOT components based on numerical considerations with positive and negative factors. You can use statistics, hypothesis testing, and numerical data collected in measurement phase. If you have a scale of values for all SWOT components, then you can use weighted average for analyzing the impact of system entities.
2. Determine the impact of SWOT analysis of the overall output based on numerical considerations. The following 3D graph can help to visualize the impact of Thread-Opportunity and Weakness-Strength impact on the output. The output would excel in the negative direction as the threads and weaknesses in the system move towards the negative direction. Similarly, the output would have a positive trend if the strengths and opportunities for improvement in the system increased in a positive direction. The quantification can help to visualize the impact of SWOT on the output as pointed out by the arrows on the image below.

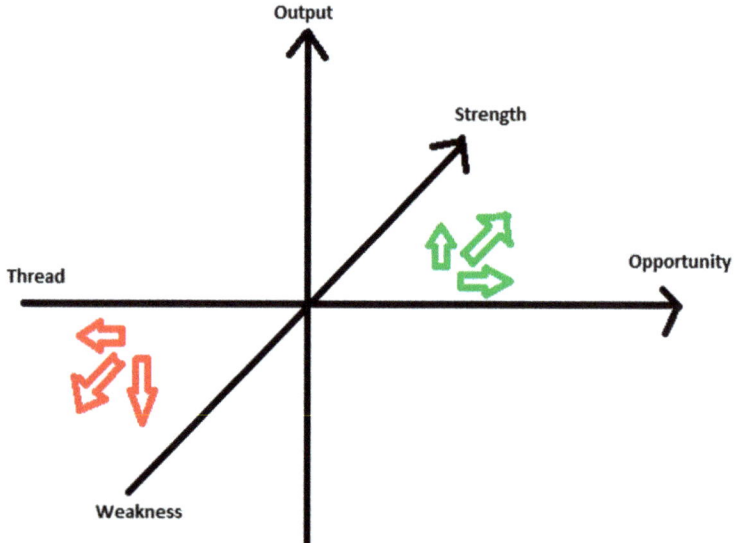

3. Assess whether the SWOT components are internal or external. Label internal factors with I and external factors with E. You can further observe by splitting these groups and identifying their impact on the overall processes and systems that relate to the defined problem.
4. List and summarize SWOT related findings in categories that are observed.
5. Design an improvement strategy based on the findings of the SWOT table.

Example (*Security Tool Coverage*). Revisiting the *Security Tools' Coverage* example, the strength of the system does not need DLP tool while the weakness is the poor utilization of the remaining 7 tools. There is 100% opportunity for improvement noting that there is 100% mismatch of tools and targets. The thread is increasing with the number of tool-inventory mismatches over the three months with a quadratic regression increase. There is an opportunity to decline this quadratic increase by internal resources through careful analysis of the system.

Example Revisiting the *Vulnerability Patching* example, the following can be attained:

- **Production**: 6th and 7th months are the outstanding months from production perspective in which the values that exceed one standard deviation (1SD) range as observed. Production over 30 only exceeds 1SD range during the first month which is something that should be investigated. There is only one data point which is out of 6 sigma range which occurs during the 6th month for production over 30. There appears to be too much variation in the data set indicating inconsistency in handling issues.
- **Development**: 2nd and 3rd months are the outstanding months from the development perspective in which the values that exceed 1SD range are observed. These months need to be observed to determine what was done differently. There is only one data point which is out of 6 sigma range which occurs on the 6th month for development under 30. There appears to be too much variation in the data set indicating inconsistency in the treatments.

6.19 Time Analysis

The time analysis of a system can solely focuses on the time aspect of the project and nothing else; therefore, other important aspects such as money, productivity, human power etc. might need to be ignored initially. To understand and not exceed other costs, the six-sigma team also needs to keep track of the costs associated with other variables. The goal of the time analysis is to focus on the time variable in the project. In some projects, time is the most crucial aspect that the customer cares about. For instance, the detection and prevention of cyber thread is expected to be done in a timely manner and time is the most important variable. If there is a security breach, the results can be dramatic, and the customer doesn't want such a breach. Another example is meeting customer deadlines to prevent penalty costs because of late delivery to the customer. In some cases, the time analysis can be completely irrelevant to the project. For instance, the quality improvement in a healthcare system that already meets the existing healthcare criteria. There are certain steps that the six-sigma team need to follow for ensuring correct time analysis:

- Is the measured data containing all the aspects of time analysis? You may have to implement end goal analysis along with the time analysis.

- What are the variables associated with the time analysis variables and is there data collected on them?
- Is the customer consulted on the level of expectations on the time analysis? Checking with the customer is always an essential part of the work to be done.
- Is the system remaining unchanged as the project progresses? It is sometimes the case that the system in place can alter from the initially started phase.

Example (*Security Tool Coverage*). The following two images outline the virtual and no-virtual options when the target needs the inventory however the corresponding items are not existing as tools. For the 7 options CMDP, EDR, HIDS, MFA, NIPS, SIEM, and TVM. The time series chart for the virtual option helps us to see the overwhelming increase for the need of tools 1 to 4 during June while there wasn't much change in the need of tools for June and July.

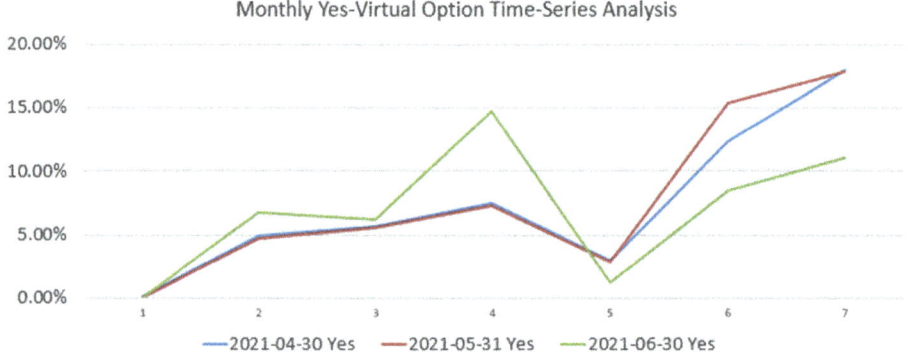

In the following figure, similar to the virtual time series chart explanation, tools 1 to 4 are needed the most in June more than April and May. Tools 6 and 7 are also more needed in June than April and May. There is not much change of needs in tools for the months April and May for any one of the tools.

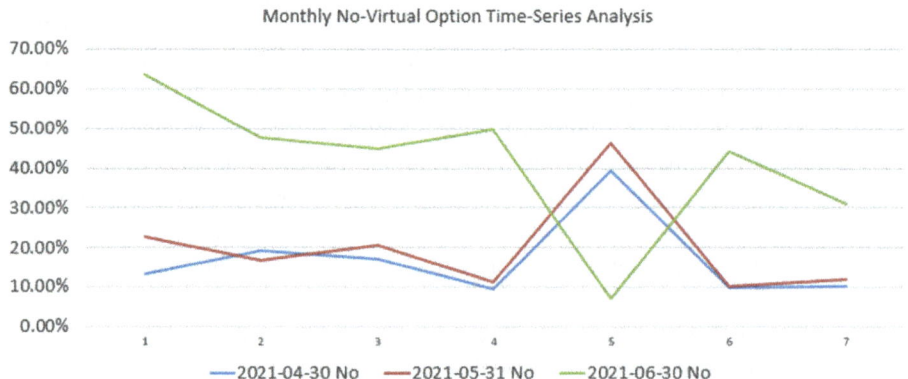

6.20 Value Analysis

In some projects the most important consideration of the project can be the "value" analysis of the project. The value in this case can be VA, BVA and NVA analysis by using the measured data and identifying where they exist in the system. These three important aspects' analysis in the project would help to outline the three layers of value. Noting that determination of NVA is the target of a six-sigma project, wastes are determined by using the value analysis. It is a simple percentage calculation for determining the value analysis for all three processes:

NVA% = (total time of NVA)/(total time of work done).
BVA% = (total time of BVA)/(total time of work done).
VA% = (total time of VA)/(total time of work done).

This analysis can help with measurable outcomes on pre-improvement phase. The value analysis worked out after the Improve phase can specify the measurable changes in the system. For instance, if the wasted work done is determined to be 12 h in a total of 120 h, the NVA% is 10%. Suppose improvements are determined, applied and the data is recollected to determine if the improvement worked. It is determined that the wasted work hours are reduced to 4 h which translates to $4/112 = 3.57\%$. This is a clear indicator that the system's capability has increased from 90% to 96.43%. Continuous improvement can be applied to reduce the waste further. In this case, it is not only the case that the %VA of the operations increased, the cost of the product is also decreased for the customer because of the reduced waste.

6.21 Value Stream Mapping (Revisited)

Using and updating the Value Stream Map that we previously introduced is one of the fundamentally important applications of a Six Sigma project when it is possible to design such a map. Continuous update of the map during Measure, Analyze, Improve, and Control phases is possible with the measured quantities needed for the project, analysis results attained during the Analyze phase, improvements determined during the Improve phase, and sustainability needs outlined in the Control phase on the map. The update of VSM is particularly important during the progress of a Six Sigma project to visualize interrelated entities of a system along with the reflection of their interactions on the map. After identification of the improvement opportunities specific to the project, the Six Sigma team needs to mark these opportunities in their respected locations on the map with a brief explanation of it. Examples of such improvements are marked in the figure below in three different locations of the system that relate to the following:

- The supplier's delivery time can be improved.
- There can be an improvement during the transition period of storage to printing
- An improvement opportunity during the transition from cutting to printing exists.

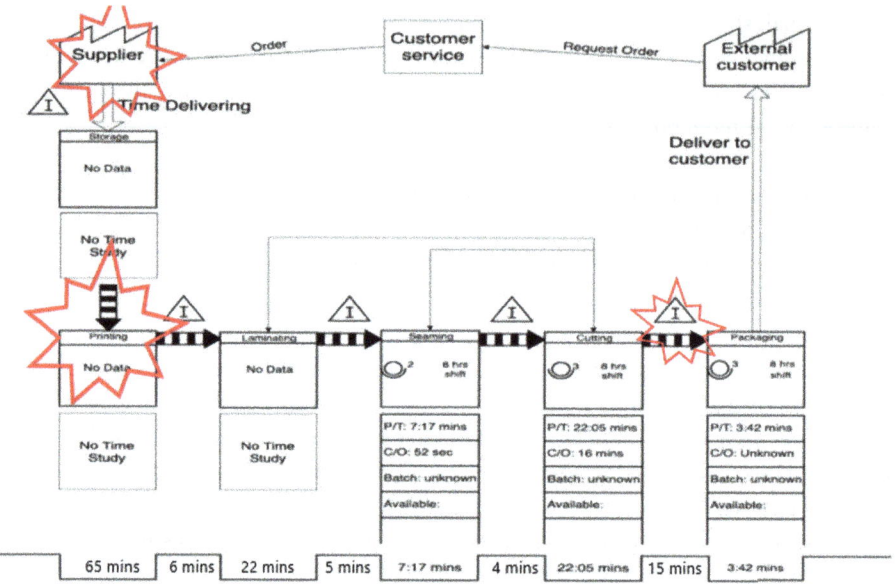

6.22 Regression Analysis

Regression analysis relates to finding the best fitting regression model to an existing data in its linear form along with the correlation analysis. We focused on the correlation analysis previously while regression analysis relates to finding the best fitting model to a given data, It is essential to note that a linear model may not be necessarily the best fit to the data. Technology can be used for complicated data sets based on personal preferences.

Example (*Security Tool Coverage*). . The following demonstrates the percentage of tools needed for the targeted cybersecurity coverage however such an inventory does not exist. The following bar chart complements this information.

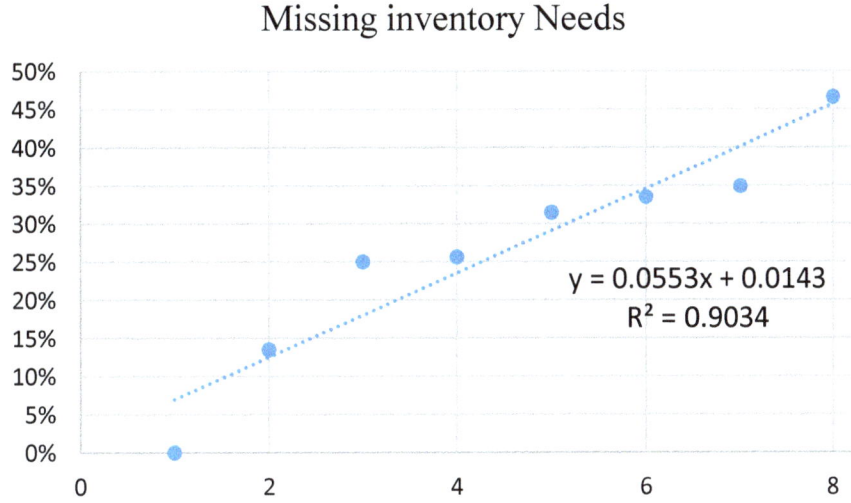

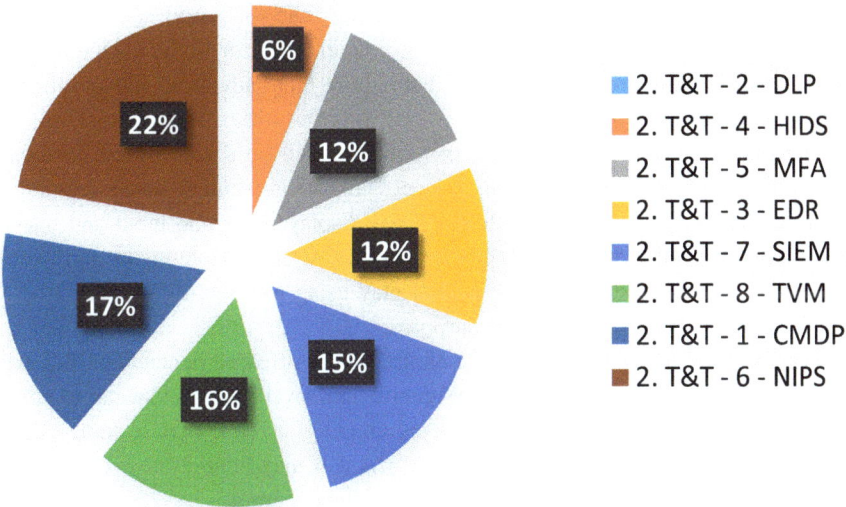

6.23 Analyze—Software

There are many different aspects of software that we need to consider in a six-sigma project. The evolution of software as computing tools for six-sigma projects has been developing dramatically since 2000. Some of the software packages used for six-sigma applications that were available as early as 2011 such as STATISTICA and Telelogic System Architect are bought out by other companies and there are software resources that still exist such as R software [27], Minitab, SigmaXL, Microsoft, and IBM products. In addition to the software packages that are needed for implementing statistical analysis for six-sigma applications, there is software needed for creating visuals, data management platforms, and implementing advanced analytics etc. There are factors that also impact the choice of the software such as size of the data, the type of the data, customers' desired software use (if required), and the expected outcome from the project. The growth of enterprises that started collecting and storing more data resulted in the need for more sophisticated methods for data analysis in six-sigma projects. The improvement of technology in areas such as robotics, hardware data storage, and cybersecurity increased the need for more in-depth analysis for six-sigma projects. The capabilities and limitations of the software along with the capabilities and budget of the six-sigma team also impact on the outcomes. Therefore, the choice of a software for six-sigma projects depends on multiple factors:

- Client's software choice. If the client has a specific software choice, then the six-sigma team either adopts someone with the corresponding knowledge or works with a consulting company/individual for development of the software.
- Size of the data: The size of the data is an important factor in the choice of software. For instance, there are large data sets that require more advanced software than Excel for analysis and visualization.
- Mathematical rigor: The choice of the software will be driven by the mathematical rigor of the project. For instance, it is possible to implement linear regression and correlation analysis in Excel with R^2 values displayed for one input and one output (by using "legend") on a graph however correlation of multiple inputs and one output may be much easier to determine and display using Minitab, SPSS, MATLAB, JMP, or R as a part of cross correlation purposes.
- Expected level of data analysis: The choice of the software depends on the expected level of analysis. Some projects might require light level analysis that can be completed by using only Excel or basic level of programming in Python while some other more complicated projects may have large data sets that may require the six-sigma team to use high level analysis techniques such as machine learning.

For a comprehensive six-sigma project, a combination of the factors mentioned above may need to be essential to apply therefore there can be multiple software packages used in a project. There are also powerful programming languages that are not commonly known to be used for Six Sigma projects or started to be used recently [1]. Noting that customer satisfaction is the key to a successful Six Sigma project, the six-sigma team needs to consult the customer on the software to be used for analysis purposes.

6.23 Analyze—Software

The rest of the section is going to cover the capabilities and limitations of software packages that can be used in six-sigma projects' data analysis as well as useful web resources for quality and six-sigma related elements' design.

6.23.1 American Society for Quality (ASQ)

The American Society for Quality (ASQ) is a knowledge-based global community of quality professionals that provides a broad range of quality related training, quality tools and principles to support working professionals and students. There are quality related tools that are shared with the general community that include lean six-sigma components. It is a very comprehensive resource for quality inspection in many different aspects that includes articles, books, case studies, training, and videos. There is an extensive list of lean six-sigma related Excel templates and information shared with the public for analysis purposes in this website under Quality Tools link of Quality-Resources section including the following [2]:

- Charts (such as Control Charts, flowchart, Gantt chart, process decision program chart, Pareto chart, Histogram etc.)
- Diagrams (such as Fishbone Diagram, Plan-Do-Check-Act (PDCA) Diagram, Scatter diagram, Box and Whisker, etc.)
- GR&R
- DOE
- FMEA
- Flowcharts
- Stratification diagram (an Excel template used for analysis of data collected from various sources to reveal patterns or relationships often missed by other data analysis techniques. Data sets can be viewed independently or in correlation to other data sets by using unique symbols for each source.)

ASQ resources target increasing the general six-sigma knowledge of viewers and provide Excel tools that can be used for small data sets; therefore, the software tools provided cannot be used for large data sets. The software to be mentioned throughout this chapter will be covering the software that can be used for analysis of large data sets.

6.23.2 C

C is a low-level programming language that requires plenty of work to be done if you want to do data analysis in a six-sigma project. It takes a long time to be able to accomplish that require high-level computing with this software. It is the foundational language for many other languages such as Perl, Python, C#, and C++. It might be a better idea to develop Six Sigma solutions in object-oriented programming languages and some other software that will be explained in this chapter [3]. There are statistical software packages such as R and SAS that got recently integrated into C and this can help with Six Sigma projects for those who are familiar with the C programming language.

6.23.3 C++

In the Measure phase we covered the file formats so that C++ can read the data. C++ is a strong programming language with the ability to handle large data sets' analysis. One valuable aspect of C++ is the unchanging nature of the software; the libraries and packages introduced by developers throughout the years are still available regardless of the version. One other advantage of using C++ is its speed in analysis of data sets. A reason for using C++ as a part of a six-sigma project would be a reflection of software's popularity on the developed libraries that can be used and its ability to handle large data sets in real time. Another reason is its technical strength to handle large data sets along with its use for complex machine learning algorithms and/or ensemble algorithms. C++ can handle large data sets with several terabyte-petabyte and implement calculations in a reasonable amount of time. However parallel processing techniques or cluster computing my be often required.

Therefore, there are many users of C C++ who developed libraries that can be used as a part of six-sigma. An example is ALGLIB which is a cross-platform numerical analysis and data processing library. Programming languages such as C C++ and C#, and several operating systems (Windows and POSIX, including Linux) are eligible for using this library. ALGLIB features include [4].

- Data analysis (classification/regression, statistics)
- Optimization and nonlinear solvers
- Interpolation and linear/nonlinear least-squares fitting
- Linear algebra (direct algorithms), direct and iterative linear solvers
- Fast Fourier Transform and many other algorithms.

If you would like to plot using C++ then plot.ly library is what you need and you can hard-code yourself for plotting. You can also use a plotting library introduced by other developers in shared platforms such as cpplot in GitHub.com [5]. A challenge with the pre-developed libraries by other users is not knowing the specifics on the development methodology and insight of the library. This is one of the reasons that many advanced software developers try to develop their own functions and libraries because they would like to know the meaning of the code used in the functions and libraries. This factor also influences the efficiency of data analysis because a predeveloped library function is as good as its developer's ability of thinking, and it may be further improved.

C++ is utilized as a source of computation for programming languages such as MATLAB, R, and Python. As is well known by C++ programmers, there are certain rules that are expected to be followed in C++ programming:

i. It is your responsibility to do the garbage collection (i.e., deleting variables after completing tasks); it doesn't do this on its own,
ii. Use of pointers correctly, and
iii. Header files of the program are comprehensive.

If you are familiar with C programming already and don't know C++, it will be ideal for you to learn C++ for six-sigma analysis purposes. We recommend using C++ for six-sigma analysis in cases when you need to

- deal with large data sets' analysis,
- use specialized algorithms for certain tasks (such as machine learning applications),
- employ a well-developed library by other users.

If you are not familiar with C++ programming, it might be a better idea to learn Java instead of Six Sigma applications if you like coding.

6.23.4 Java

The popularity of Java programming is not questionable for those who have extensive programming experience. The acceptance of Java programming among companies makes it a strong candidate for using it in six-sigma applications. The main application difference between C++ and Java program is the fact that C++ is mainly used for system programming while Java is mainly used for application programming. It is widely used in window, web-based, enterprise and mobile applications [6]. There is no big difference between C++ and Java when we compare from a six-sigma applications perspective.

Java is one of the earliest programming languages used for enterprise development therefore the long history of Java is an indicator of significant development made by many companies. Some of the important aspects of Java are the following [7]:

- Java can be used for data analysis, including data visualization, cleaning data, importing, and exporting data, statistical analysis, deep learning, and Natural Language Processing (NLP).

- It is natural to face Big Data in many applications and most of the popular frameworks and tools used for Big Data are typically written in Java. This includes Fink, Hadoop, Hive, and Spark.
- A developer can write code that is identical across multiple platforms by using the Java Virtual Machine which makes it one of the best platforms for machine learning and data science. It also allows them to create custom tools at a faster pace and features a load of IDEs that help to improve overall productivity levels.
- Lambda Expressions which is a part of Lambdas that was introduced in Java 8 simplifies the development of large enterprise projects significantly for developers.
- An important aspect of a programming language is scalability, and Java makes application scaling an easier process for programmers. This makes Java a strong candidate for the building of larger or more complex Artificial Intelligence and Machine Learning applications, especially when they are being built from scratch.
- The speed of Java makes it a strong candidate for Data Science and Machine learning applications when compared to many other programming languages. Many of the most popular websites and social applications of today such as LinkedIn, Facebook, and Twitter rely on Java for their data engineering needs.
- Production codebases are commonly written in Java. Knowing Java helps developers figure out how data is being generated, submit merge requests to production codebases, and deploy Machine Learning solutions to production.
- There are many libraries and tools developed for Java including Data Science and Machine Learning. Weka 3 is an example of a fully Java-based workbench popularly used for algorithms in machine learning, data mining, data analysis, and predictive modeling. In addition, Massive Online Analysis is open-source software used specifically for data mining on data streams in real-time.

Java programming language is useful, fast, and reliable in many different application platforms such as data mining, data analysis and machine learning. It is certainly worth using this program for analysis purposes in Six Sigma applications. In the case of the need to analyze a large data set for a six-sigma project, Java is certainly a strong tool that can be used. The graphs, figures, charts etc. can also be coded in Java for displaying the results, noting that the graphing of such data will not be possible by using Excel or similar software.

6.23.5 SPSS by IBM

SPSS is a straightforward and easy to use software with English-like command language [23]. It has an extensive and thorough user manual. This software is particularly popular in healthcare research however market researchers, survey companies, government agencies, education researchers, marketing organizations, data miners, and many others used the software for the processing and analyzing data.

SPSS meets four different needs of data analysts:

- SPSS—Statistical analysis: Basic statistical functions such as frequencies, cross tabulation, and bivariate statistics. Figures, charts, graphs etc. are also easy to implement in the Excel-like data entry environment of SPSS [8].
  - Descriptive statistics—components such as frequencies, cross tabulation, and descriptive ratio statistics etc.
  - Bivariate statistics—components such as ANOVA, mean, correlation etc. are possible to determine by using the software.
  - Correlation Analysis—linear regression.
  - Predictive analysis—clustering analysis and factor analysis.
- SPSS-Modeler: Building and validating predictive models using advanced statistical procedures is possible by using SPSS Modeler Software [9].
- SPSS—Text Analytics for Surveys Program: Helps survey administrators uncover powerful insights from responses to open ended survey questions [10].
- SPSS—Visualization Designer: Allows researchers to use a variety of visualization tools that will help them to communicate their analysis results anywhere from basic charts and graphs to advanced graphs. Advantages of the software include:
  - built-in visualization templates,
  - "Drag-and-drop" graph creation,
  - Ability to work with data sources such as IBM SPSS Statistics data files, and common database sources such as: DB2®, SQL Server™, Oracle® and Sybase® [11].

In addition, SPSS also provides solutions for data management, which allow researchers to perform case selection, create derived data, and perform file reshaping.

A metadata dictionary can be formed in SPSS for data documentation that acts as a centralized repository of information pertaining to data such as meaning, relationships to other data, origin, usage, and format.

SPSS is a software package that can be used for six-sigma project purposes depending on the scope of the project. Above-mentioned capabilities of the SPSS software draw the boundaries that SPSS can be used in a six-sigma project.

6.23.6 MATLAB

MATLAB is a high-level programming language that allows analysis in many different areas of interest including deep learning and machine learning, signal processing and communications, image and video processing, control systems, test and measurement, computational finance, and computational biology. There are tutorials and a wide range of developed tools and codes available for MATLAB users including the following [3]:

- Math and Optimization Toolboxes: Curve Fitting, Optimization, and Global Optimization, and Mapping.
- AI, Data Science, and Statistics Toolboxes: Statistics and Machine Learning, Deep Learning, Reinforcement Learning, Text Analytics, and Predictive Maintenance.
- Database Access and Reporting Toolboxes: Database, and MATLAB Report Generator.

Simulink of MathWorks is another product that can be used for analysis of specific applications (such as electrical, fluidic, etc.) There are also general-purpose applications such as *System Composer*™ that enables the definition, analysis, and specification of architectures and compositions for model-based systems engineering and software design.

MATLAB has a comprehensive list of plotting options along with Apps that allow users to employ pre-developed codes for applications. In addition to basic plotting options, the software has 2D and 3D Stem and Stair plots, bar plots, scatter plots, graph plots, pie charts, histograms, polar plots, geographic plots, contour plots, image plots, surfaces, volumetrics, vector fields, analytics plots, curve fitting plots, financial plots, geographic data plots, signal processing toolbox, statistics, and machine learning plots. The categories of Apps made available by MATLAB that can be particularly helpful in Six Sigma applications include curve fitting, optimization, machine learning and deep learning, math, statistics and optimization, and simulation graphics and reporting. There are other Apps that can be used for specific six-sigma applications such as signal processing, computational biology, and computational finance offered by MATLAB.

MATLAB can analyze large data sets and employs C and C++ in several applications. The linkage to C and C++ also makes it a strong programming language in addition to its own capabilities. A strength of the software is the support provided, and the code shared in web-based communities. In several cases, the application can be as simple as modifying the code shared by MathWorks' support to fit your application. Some of the advantages of MATLAB are the following:

- Algorithms can be implemented and tested easily.
- Apps developed for ease of applications.
- Built in algorithms available for users.
- Computational codes can be developed easily.
- Easy debugging.
- Ease of image processing and creating simulation videos easily.
- Extensive data analysis and visualization can be performed.
- External libraries can be called for use.
- Graphics user interface allows users to develop figures, graphs etc.
- Symbolic toolbox allows the symbolic computation easily.

It is therefore possible to implement a variety of Six Sigma analysis in MATLAB; however, it is essential to know the programming language to accomplish tasks. MATLAB does not contain tools specific to six-sigma such as VSM, Kano model etc.

6.23.7 Microsoft Products

There are many Microsoft Office products that are typically used in classic Six Sigma applications such as Word, PowerPoint, Access, and Excel. Microsoft Visio is a useful tool that can allow users to design flow charts. There are many features of Microsoft Access that can be used for project planning and calculations that help with management of the project. Microsoft Project Software can be used as a part of Six Sigma as well and it has advantages that one can utilize that are also related to project management. An in-depth analysis is possible by using Microsoft Excel that will be explained in the next section.

6.23.8 Microsoft Excel

Excel is one of the primary data collection platforms that companies use in the World. It is not only used for data collection but also analysis of data sets. There is a wide range of tasks that can be accomplished with data stored in Excel. There are built-in functions of the software that are popularly used for analysis of data stored in Excel. Functions such as Average(), STDV.S() (standard deviation for a sample set), STDV.P() (Standard deviation for a population), and COUNT() are popular functions used for data analysis. Using these functions is as simple as tying " = " initially in an Excel cell and then choosing the range of data. For instance, typing the expression $= AVERAGE(A1:A10)$ in a cell calculates the average of data stored in A1 through A10 cells. There are many features of Excel that help with data analysis:

- Insert tab: Contains pivot tables, charts, figures, tables etc. Pivot tables can be particularly helpful in analysis summary of data sets in many ways.
- Formulas tab: Helps you to find the formula that you are looking for based on categories such as Finance, Math and Trigonometry, Text etc.
- Data tab: Contains *What-If Analysis*, *Forecasting sheet*, data transfer from a variety of resources (such as files, databases, web etc.)
- View tab: Contains *Macros* that allows users to enter their own Excel code or record an action in Excel that would generate the VBA code for the user. More information on VBA will be covered in the next section.

Our project experience with companies showed the extensive use of Excel in either storage or analysis of data sets in Six Sigma applications. The project data was usually stored in Excel files and the analysis of the data was expected to be done by either using Excel or some other advanced programming language such as Java, MATLAB or Python if the data set is large. Some of the most useful components of Excel in analysis applications are listed below:

- Excel plotting—The most useful graphs for a six-sigma project include column, line, pie, bar, area, scatter, histogram, Box and Whisker,
- Excel functions—Useful for calculating average, count, standard deviation, minimum, maximum and sort.
- Pivot tables—Useful for statistical summary of large Excel data sets to implement sum, count, average, min, max, product, standard deviation, and count,
- Macros (to be explained in the next section.)
- Add-ins—Custom made Excel applications that serve a specific purpose. Add-ins appear as a separate tab in Excel, and they can be added to Excel. Some of the Palisade products that are add-ins to Excel will be explained in a later section.

The above-mentioned abilities, specifications and components of Excel outline the program's analysis capability. There are many web-based training and learning resources for Excel that can be easily learned with no technical programming background needed.

6.23.9 Visual Basic for Applications (VBA)

Visual Basic computer language was created by Microsoft in the 90's that allows Microsoft programs to communicate with each other to accomplish tasks. The language is now used in Office programs such as Excel, Word, Outlook, Access, and PowerPoint along with programs such as NotePad and Paint. Macros are blocks of VBA code that perform specific tasks. The coding created by Microsoft is very user intuitive for its applications. For instance,

```
    For Each storeSales In Range("C7:C17").
        If storeSales.Value <> "" And storeSales.Value <> 0 Then
            curSale = storeSales.Value
            countOfSales = countOfSales + 1
            sumOfSales = sumOfSales + curSale
            minStore = IIf(curSale < minSale, storeSales.Offset(, −1).Value, minStore)
            minSale = IIf(curSale < minSale, curSale, minSale)
            maxStore = IIf(curSale > maxSale, storeSales.Offset(, −1).Value, maxStore)
            maxSale = IIf(curSale > maxSale, curSale, maxSale)
            If curSale < 500 Or curSale > 5000 Then
```

```
            'capture reason for low or high sales
            reason = InputBox("Reason for deviation:", storeSales.Offset(, −1).Value)
            storeSales.Offset(, 1).Value = reason
      End If
End If
```

The use of the Excel range C7 to C17 in the first line of the above code makes this programming environment unique compared to many other software packages, noting that the data storage environment is same as the programming environment. Few other software packages such as SPSS and Minitab have similar capabilities. Noting that Excel is used by many companies in the World, it allows many users to analyze data sets with limited computer programming knowledge to easily pick up on how the VBA language works. VBA program is very similar to Python programming language in the way it is presented to the compiler to run the code (which is user friendly for the programmer to read and write the code.)

There are many analysis (and additional) tasks that can be accomplished by using VBA in Excel. Some of the very useful tools are the following:

- Macro recording: You can click on "Record Macro" available under the "Macros" section of Excel and start recording your work on the Excel cells that will generate a Macro code for your entries. For instance, suppose you have data in A1 to A10 cells of Excel and you click on Record Macros first. You go back to the spreadsheet where the data is available, and you click on cell A13 and start typing = *COUNT(A1:A10)*. If you click on "Stop Recording" under the "Macros" section, the following code will be generated under the Macros section automatically once you save it:

```
Sub Macro1()
    Macro1 Macro
        ActiveCell.Formula = "=COUNT(A1:A10)"
        Range("A13").Select
End Sub
```

- Button generation: You can create buttons on Excel that will allow you to click on and generate a solution based on the VBA coding. There can be multiple buttons created that can appear on the Excel spreadsheet that can appear right next to the data. This automates the analysis process and helps to reduce waste in the cases when repetitive analysis is implemented.
- Task Automation of Spreadsheet Analysis: There are many ways that the VBA code can apply out-of-box analysis in Excel including opening a blank spreadsheet, naming it, and storing the analyzed information on the new spreadsheet directly. It also allows the program to retrieve data from other Excel spreadsheets and resources. In many

ways, there are many different analyses that can be completed. All this can be coded in a Macro and a button can be generated which would allow the user to click on a button in the way it is explained above.
- User Interaction: Interaction with the user can be accomplished in a way that the user can respond to a set of questions that are populated as message boxes in Excel and the analysis can be presented after data entry. This also adds another dimension to the use of Excel data.
- Extensive Function Entry for Data Analysis: It is possible to enter anything in a Macro that is available in Excel including functions, figures, graphs, charts etc. for extensive data analysis. A click of a button can generate the entire expected analysis of a data set however coding can take a lot of time.
- Dynamic range: VBA coding allows the use of dynamic data range for analysis purposes. For instance, if you continue adding data into a spreadsheet and if you want to continue determining the maximum value in the data set within this dynamic data set, it is possible with the click of a button.

There are also web-based pre-developed codes and learning modules shared by VBA users and developers for you to retrieve and modify codes for your own needs. Some of the fundamental tools that you can use in your VBA code before you start learning are the following:

- **Subroutine (Sub)**—Used as the opening statement for the Macro code that should be ended with the statement "End Sub". A new macro or procedure is created every time a Sub is created.
- **Module**—Functions and macro codes are written under modules. Recording and storage of Macros happens in Modules.
- **Class Module**—Your own custom classes, methods, and collections into the VBA library can be added in Class Module. It is a powerful tool for advanced programmers.
- **Function**—You can create your own custom functions in VBA that can either be used as a part of macros with a certain purpose or they can be used for a certain task that can be placed in the Excel Formula Bar.
- **Userforms**—Entering inputs and choosing options as a part of the VBA implementation is possible by using Userforms. Pop-up boxes can be generated by Userforms that allow users to enter inputs or choose options such as user input boxes, Error Message Boxes, and Dialog Boxes. You can also create your own custom Userforms.

The use of VBA in Excel is very powerful and helpful in many aspects, however you always need to keep in mind the limitations on data storage of Excel. It is possible that Excel can crash while trying to handle the code in a Macro for a large data set, therefore it is best to use Macros in Excel for relatively small data sets' analysis.

6.23.10 Add-Ins

Microsoft strengths the analysis ability of Excel through a program called "Add-in" that can be attached to Excel to give it additional functionality. There are add-ins readily available to be used in Microsoft Excel such as "Analysis Toolpak" as well as other add-ins that can be downloaded and used such as StatProGo. Analysis ToolPak allows you to implement data analysis including ANOVA, correlation, covariance, descriptive statistics, regression, Fourier analysis, and some other basic statistics as well as statistical tests such as t-test, z-test and F-test. Once an add-in is installed, it can be used as a part of any Excel workbook. The extension of an Excel add-in is ".xlam" and it is a workbook that Excel can open automatically when it starts up. It can be found under the Data Analysis option of Data tab in Microsoft Excel 2020. An Excel add-in is very user friendly and easy to use for small data sets; however, the data-size limitation of add-ins, as previously mentioned, is limited to the Excel spreadsheet's size.

6.23.11 Minitab

Most Black Belts and Master Black Belts favor Minitab® Statistical Software. This package has been around for many years and is a favorite of universities and colleges teaching statistics. Minitab is a very comprehensive statistical analysis package designed for serious statistical analysis. Don't try it at home without some serious training as part of an Advanced Green Belt or full Black Belt course.

6.23.12 Palisade Products

Palisade company offers a variety of Excel-based products that can help with six-sigma analysis and a variety of applications. Each one of the Palisade products listed in this section contains useful example models, expert videos, available training, and custom solutions. The support by the company includes videos for common questions, searchable knowledgebase, webinars, and requesting help. The products offered as a part of DecisionTools Suite by Palisade to be explained in this section include the 8th versions of the following Add-ins:

- @Risk
- Evolver
- Neural Tools
- Precision Tree
- StatTools
- TopRank.

Noting that all the above listed Palisade products are Excel Add-ins, the analysis capability of the software will be limited to the Excel spreadsheet information.

6.23.12.1 @Risk

Add-ins of Excel can be powerful when they provide an extensive number of statistical packages and many other analysis features and @Risk is one of these add-ins. Many of these useful tools include a variety of analysis applications including the following:

- **Distribution and Fit tabs**: A variety of continuous and discrete distributions can be generated either automatically or fitting a distribution to a given data set.
- **Function tab:** Contains a variety of functions written as Macros. There are three main sections of the function tab: Distribution Functions, Statistical Functions and Other Functions. Six Sigma is an option listed under Statistical Functions with the focus on risk-based functions.
- **Correlation tab:** Allows users to enter the correlation input data from Excel cells. Repeated Correlation, Time Series Correlation, and Copula based methods are shared as a part of the Correlation tab.
- **Simulation tab:** Options in this tab include Excel simulation, Goal Seek analysis, Stress Analysis and Advanced Sensitivity Analysis.
- **Optimization tab:** In the case of the need for solving a problem that requires an optimization problem that is either mathematically formulated or based on the Excel spreadsheet data, you can rely on the optimization tab.

@Risk can be mainly used for Excel based information therefore the analysis capability of the software will be limited to the Excel spreadsheet information.

6.23.12.2 Evolver

Evolver is a software package that can be used when the analysis of the collected data for a six-sigma project requires an optimization application. The software can solve a wide range of optimization problems. The software employs a genetic algorithm that was launched in 1989.

6.23.12.3 NeuralTools

The Neural Network analysis in a six-sigma project can be implemented by using NeuralTools add-in of Palisade. Training the network on your data, testing the network for accuracy, and making predictions from new data are the three basic steps in a Neural Networks analysis. These tasks can be accomplished by NeuralTools automatically in one simple step. The predictions are updated by NeuralTools automatically when the input data changes, so you do not have to manually re-run predictions when you get new data. It is possible to combine Excel's Solver with Palisade's Evolver to optimize tough decisions and achieve goals. It contains click and choose items such as Data Set Manager and

Viewer along with the Train, Test and Predict options for Neural Network implementation. Some of the advantages of the Add-in include the ability to automatically update predictions when data changes, making predictions with incomplete data, and using both categorical and numerical data.

6.23.12.4 PrecisionTree

PrecisionTree [29] add-in by Palisade allows users to choose from a variety of analysis options that can be helpful in a six-sigma project. Decision trees are quantitative diagrams with nodes and branches representing different possible decision paths and chance events. PrecisionTree uses Microsoft Excel data to address complex, sequential decisions by visually mapping out, organizing, and analyzing decisions using decision trees. The analysis of the data can help you to identify and calculate the value of all possible alternatives for you to choose from a variety of alternatives. Some of the tasks that can be completed include decision tree construction, risk and sensitivity analysis, Bayesian Regression and determining Model Errors.

6.23.12.5 StatTools

StatTools add-in by Palisade might be what you are looking for if you are interested in an Excel based statistical analysis tool. The software contains many features that can be utilized:

- **Data Utilities**: Allows users to stack, unstack, transform, and lag data. Interaction between two variables, random number generation, and combination of variables are also possible using this tab.
- **Statistics**: One variable statistical summary, correlation, covariance, confidence interval, sample size selection, hypothesis testing, one-way ANOVA, two-way ANOVA, chi-square tests, normality tests, regression analysis, logistic regression, nonparametric tests, and discriminant analysis are possible statistical calculations.
- **Quality Control**: Pareto chart, X/R charts, P-chart, C-chart, U-chart are possible quality charts that can be constructed.
- **Multivariate Analysis**: Principle component analysis and cluster analysis are possible to implement.

Like all other Palisade products, PrecisionTree allows you to take advantage of data entry in Excel while it cannot allow large data set analysis due to its connection to Excel.

6.23.12.6 TopRank

TopRank by Palisade [30] is designed to apply what-if analysis to determine the role of the input and its impact on the output through the logic "**what** happens to the output **if** we make changes on the decision that impact the input". It helps with the analysis of critical factors that play a role in efficiency analysis. The user can define any output cell or cells, and TopRank is capable of automatically finding and varying all input cells which affect the output. It is also possible to control the cells that need to be analyzed for the six-sigma project. Similar to the other add-ins developed by Palisade, the software is powerful to be used for six-sigma analysis applications while it is limited to Excel.

6.23.13 Perl

Perl is an open-source program that supports object-oriented, procedural, and functional programming. Perl includes powerful tools for processing text that make it ideal for working with HTML, XML, and all other mark-up and natural languages. The software interfaces with external C/C++ libraries through XS or SWIG. Embedding Perl interpreter into other systems such as web servers and database servers is possible [32].

Perl has been utilized by a limited number of companies for Six Sigma purposes, and it has strong potential for analysis purposes, however it has not been utilized in research settings. There are no books or textbooks on the applications or use of Perl for Six Sigma purposes.

6.23.14 Python

Python is one of the most popular object-oriented software with increasing interest of users for analysis purposes. I remember one of my computer science students stating "I don't know why anyone would want to choose using a language other than Python because what you type is what the program does…" It is a very user-friendly programming language with many packages developed for users. Python is popular among programmers therefore much focus and effort are put into developing packages. For instance, there are packages developed for machine learning and there is even a top 10 list with the top three options being

- #1: TensorFlow
- #2: Pytorch
- #3: SciPy.

It is a strong programming language that can handle large data sets and be useful in complicated data sets analysis. As one can imagine, it takes a good amount of coding to sketch a graph or figure when compared to software such as Excel, SPSS, and Minitab. The strength of Python is its "user friendly" script that allows ease of programming, although one still needs to be very comfortable with algorithm development and the corresponding coding to accomplish complicated tasks. For instance, the following code for opening a file in Python is easy to read for a person who is not familiar with coding:

Try:
    F = open('myfile.fa")
Except:
    Print("File myfile.fa does not exist!")

Some of the additional powerful packages in Python include.

1. **Faker**: Used for creating fake and random data.
2. **NumPy**: Used for analyzing numerical data and provides sophisticated functions in n-dimensional array objects.
3. **SciPy**: Used for many different purposes including statistical analysis, image processing, I/O operations etc.
4. **Matplotlib**: Used for generating visualization tools such as histograms, charts, and plots that help with 2D plotting and data analysis by using visuals.
5. **Nmap**: Used for network discovery. Available hosts in a network by using raw IP packets can be determined by using Nmap along with the services (application name and version) that the hosts are offering, the operating systems (and OS versions) they are running, and the type of packet filters/firewalls they are using.
6. **Pandas**: Pandas is a popular package used for machine learning and data analysis by which training, modeling and analysis of machine learning tasks are possible [12].
7. **Cryptography**: Common cryptography algorithms take place in this package that range from high level to low level. The idea of the implementation is based on Fernet; a crypted message cannot be read or changed without the key (or the "secret key").

These packages can be useful in six-sigma implementations and useful for projects. There are many other packages that can be utilized as a part of Python that can be found on the web.

The use of Python would be very easy for six-sigma applications if you are comfortable with the script. It would be ideal to use Python on projects that have large data sets. Recently, Python and R software (to be covered in the next section) are integrated to take advantage of both platforms. This integration can allow you to do more sophisticated analysis, however you would need to learn the integration of the platforms.

6.23.15 R and R-Studio

R is a very popular (free of charge) software used for many different applications by many companies and software developers. R-studio is the university edition of R developed for educators. It is a platform that integrates the compiler of the software to view next to the code. There is no difference between R and R-studio for programming and script purposes.

R is popular due to its statistical capabilities, ability to handle large data sets, and the packages and libraries developed for the software. It is a script-based language, and the script is not as optimized as some other languages such as Python; it requires an extensive amount of typing to accomplish tasks. However, there are packages developed for Six Sigma applications by R users (see for example [12]) and there are books written on Six Sigma applications by using R software (see for example [13]) that makes the use of R for those who are not familiar with the program. Some useful R information that can be utilized for six-sigma include the following:

- ss.cc—Plotting control charts
- ss.ceDiag—Cause and effect diagram
- ss.pMap—Process Mapping
- ss.ci—Confidence interval for the mean
- ss.study.ca—Plots a Histogram with density lines about the data of a process. Checks normality with qqplot and normality tests. Shows the Specification Limits and the Capability Indices.
- ss.rr—Performs Gage R&R analysis for the assessment of the measurement system of a process.

There are also available free resources for those who would like to learn R (see for example [14] for a comprehensive list of free resources as well as [15] for a particular resource with educational components.)

As mentioned in section that covers Python, R is recently integrated into Python that makes these two powerful software packages more powerful through the integration. There are web-based support blogs that contain many resources such as bug reporting, developer pages, blogs, manuals, certification and books that relate to R for R programmers that use Unix, Windows, and MacOS.

As pointed out above, R is a high-level language, and the expressive nature of the language impacts its speed; Integrating R to programs such as C and C++ that are low-level compiled languages can eliminate the issue with the speed of the R code. While C and C++ often require more lines of code (and more careful thought) to solve the same problem, they can be orders of magnitude faster than R [16]. Linking R to C++ starts with the code devtools::use_crpp().

It is also possible to integrate R code into Java, however it would require more advanced coding skills and some work for it to be done. Please see for instance [17] that provides integration of R code into Java.

R is certainly a very strong software if you are willing to undertake the code to written for analysis of what is measured in a six-sigma project. It is also possible to implement high level tasks using R such as machine learning (see for example [18]) however you need to undertake the expected code writing in order to accomplish tasks.

6.23.16 SAS and SAS-Studio

Statistical Analysis System (SAS) [24, 25] is a software package that is commonly used for statistical analysis. Just like script-based programs such as R, C++, Java, SAS requires coding in the corresponding software platform. There are also well-established resources for users of the software for many different applications such as six-sigma [19]. There are handy SAS procedures that help with the six-sigma tasks to be accomplished as a part of the analysis phase:

- Determining the normal distribution nature—Is the data approximately normally distributed? SAS Procedures: Chart, Rank, Plot, Means, plus Data step programming.
- Correlation analysis: SAS procedures Plot and Corr.
- Box plotting: SAS procedure Boxplot.
- Extensive statistical analysis summary: SAS procedure Univariate (which determines basic statistical measures as well as sophisticated tests that are used for location and normality.)
- Expected variation analysis based on ± 3 Sigma range: SAS procedures mean and plus DATA step programming are needed.
- Null hypothesis: SAS procedure TTEST is needed.
- Frequency analysis: SAS procedure Freq is needed.
- Analysis of Variance: SAS procedure ANOVA is needed.

The above-mentioned SAS procedures make the six-sigma experience much easier once they are comprehended by the six-sigma team.

Like the R program, it is also possible to integrate SAS with C, C++, and Java. The evolution of the SAS/C and SAS/C++ Compiler continues to reflect the development requirements of the SAS/C and SAS/C++ community with the release of SAS 7.0 [20]. The application programming interfaces provided with SAS Integration Technologies enable the developers to use Java-based distributed applications that are integrated with the SAS platform as a part of SAS version 9.1 release. The Support of SAS is very helpful for comprehending the integration of the SAS program to other software packages [21].

SAS can be used for analysis of large data sets as a part of six-sigma. Its integration into many other programming languages along with the SAS procedures makes it a strong candidate for Six Sigma projects as long as you are comfortable with the script of the software.

6.23.17 SigmaXL

SigmaXL is an add-in tool developed for Microsoft Excel similar to Palisade products [28]. It has a variety of graphic and statistical tools that can be utilized with data entered in Excel. It has many components that can be used as a part of a six-sigma project:

- **Charts**: Analysis of Means (ANOM) Charts, Pivot Charts, Pareto Charts, Run Charts, Multi-Variable Charts, Control charts.
- **Plots**: Dot plots, Box plots, Scatter Plots, Normal Probability Plots.
- **Statistical Tests**: Random number generation for statistical distributions, Chi-square and Fisher's Exact tests, Equal Variance tests, Equivalence Tests, One and Two Sample t-Tests, Nonparametric and Nonparametric Exact tests, Paired t-Tests,
- **Other Statistical Tools**: Descriptive Statistics, Correlation Analysis, Logistic Regression, Multiple Linear Regression, Minimum Sample Size for Robust t-Tests and ANOVA, One-Way and Two-Way ANOVA, Power & Sample Size Calculators, Tolerance Interval Calculator (Normal Exact).

In addition, the following templates are included by SigmaXL that can be used in a six-sigma project [31]:

- **DMAIC Basics**: Team/Project Charter, SIPOC Diagram, Flowchart Toolbar, Data Measurement Plan, Cause and Effect (Fishbone) Diagram and Quick Template, Cause and Effect (XY) Matrix with Pareto, Failure Mode and Effects Analysis (FMEA) with RPN Sort, Quality Function Deployment (QFD), and Control Plan.
- **Lean Basics**: Takt Time Calculator, Value Analysis/Process Load Balance, and Value Stream Mapping.
- **Probability Distribution Calculators**: Normal, Lognormal, Exponential, Weibull, Binomial, Poisson, and Hypergeometric distributions.
- **Other Tools**. Design of Experiment (DoE) templates, Measurement System Analysis (MSA) templates, Control Chart templates, Gage R&R, and many other graphical tools.

On the one hand, SigmaXL offers a wide variety of six-sigma tools for analysis purposes, on the other hand, similar to the other Excel add-in software, SigmaXL is limited by its connection to Excel for data analysis purposes.

6.23.18 Other Software Packages

- ARIS Six Sigma
- JMP [37]: JMP is a statistical software by SAS with many different options for users. Options include JMP, JMP live, JMP Pro, JMP Clinical, and JMP Genomics.

6.23 Analyze—Software

JMP allows users to implement Data Acquisition, Data Cleanup, Data Visualization, Basic Data Analysis (such as generating histograms, regression, distribution fitting and other analysis tools to launch data exploration.), Text Exploration; Group, Filter and Subset Data, DoE, Statistical Modeling, What-if Analysis, Reliability Analysis, and Quality and Process Engineering. JMP interface can be used to leverage other analytics tools, such as SAS®, MATLAB, R and Python. It is possible to develop applications by using JMP and there are developers who already developed six-sigma apps for JMP [38].

- LQATS—Lyons Quality Audit Tracking System by Lyons Information Systems
- *NCSS Statistical Software* [34]: NCSS is another strong statistical software package that can be utilized for statistical analysis purposes. There is a wide range of procedures that can be applied on data that can be transformed into spreadsheet and database formats and exported in the same fashion. Some of the procedures include ANOVA, a variety of charts and graphs, correlation and curve fitting, descriptive statistics, DoE, a wide range of statistical tests, forecasting, quality control, regression analysis, reliability analysis, and time series analysis. The list of statistical analysis tools and graphing capabilities of the software are very broad compared to many other software packages.
- QPR ProcessGuide by QPR Software
- *Statistical Design Institute (SDI) Tools.* SDI offers two Microsoft Excel based tools to be used for six-sigma application purposes: Triptych and Apogee [32]. In addition to many other analysis tools Apogee, some of the most recent advancements in data analysis capabilities of the software included Monte Carlo, Multi-Objective Optimization, Allocation, and Sensitivity Analysis [33]. Triptych applications focus on decision making tools and risk-based analysis.
- Software AG webMethods BPM Suite
- Statgraphics [35]: Just like the NCSS software, Statgraphics is a software that has high level of statistical capabilities; however, it is limited to the analysis of the data stored in Excel as it is software used for Excel application purposes. There are many statistical tests, control charts, graphing modalities, and general statistical analysis tools along with the six-sigma specific tools. For instance, *Sigma express* **by** Statgraphics has a different formatting for Six Sigma applications than many other software packages: The software has five options in Excel that the users can visit and use the corresponding DMAIC component. For instance, under the Define option, you can find QFD Matrix, Cost of Quality Trend Analysis, Process Mapping, and Affinity Diagram. Therefore, the tools for DMAIC are chosen by the company as a part of the software. Statgraphics also offers Centurion 19 as a stand-alone Windows application that includes a broad range of advanced statistical procedures. It is designed for master and advanced black belts who need a large set of statistical tools. Data may be input from Excel, databases, and a variety of other software programs [36].

Chapter 6 Exercises

Exercise 6.1 There is a variety of concepts explained in this section for cybersecurity applications. Which one of these concepts may be more important than the all the other concepts explained in this chapter. Please explain your response.

Exercise 6.2 There are three cybersecurity examples given in this section for analysis purposes. Which one of these is more important than the others. Given your response to Exercise 6.1, what is the outcome of the application of this concept on the example? Please explain.

Exercise 6.3 Python is one of the software packages that gained popularity in recent years. What type of other software packages can be integrated with Python that can be used in cybersecurity applications? Give three examples of such software packages of different uses.

Exercise 6.4 Optimization is another area of interest in cybersecurity applications and there is associated software that can be utilized. Find an optimization package that can be used in cybersecurity applications and give two examples of such applications.

References

1. Lopez, E., Redchuk, A., & Moguerza, J. M. R: A language and environment for statistical computing. In *R foundation for statistical computing*. Six Sigma Tools for Quality and Process Improvement, SixSigma.
2. Matlab by Mathworks. https://www.mathworks.com/discovery/what-is-matlab.html
3. Bretheim, D. R. *SAS for Six Sigma—An introduction*. Towers Watson. https://support.sas.com/resources/papers/proceedings15/2984-2015.pdf
4. Quality Tools, American Society of Quality. https://asq.org/quality-resources/quality-tools
5. SPSS Statistics by IBM. https://www.ibm.com/products/spss-statistics
6. Control Charts, American Society of Quality. https://asq.org/quality-resources/control-chart
7. Javapoint. https://www.javatpoint.com/
8. Liu, S. (2020). *Python for Six Sigma*.
9. Cano, E. L., Moguerza, J. M., & Redchuk, A. (2012). *Six Sigma with R: Statistical engineering for process improvement* (Vol. 36). Springer Science & Business Media.
10. R program. https://www.rdocumentation.org/packages/SixSigma/versions/0.9-52/topics/ss.cc
11. Lantz, B. (2013). *Machine learning with R*. Packt publishing.
12. Scrucca, L. qcc: Quality Control Charts.
13. Recchia, D. R., Barbosa, E. P., de Jesus Goncalves, E. IQCC: Improved quality control charts.
14. PERL. https://www.perl.org/about.html
15. Microsoft Office. https://support.microsoft.com/en-us/office/load-the-analysis-toolpak-in-excel-6a63e598-cd6d-42e3-9317-6b40ba1a66b4

16. JMP. https://www.jmp.com/en_us/home.html
17. Toyota Production System, Taiichi Ohno, Productivity Press.
18. Shimbun, N. K. (1989). *Poka-yoke: Improving product quality by preventing defects*. CRC Press.
19. Tokgöz, E. (2024). *Quality & Lean Six Sigma applications for industrial engineers*. Springer Nature. https://link.springer.com/book/9783031557392
20. Tokgöz, E. (2024). Quality and Lean Six Sigma for engineering technicians, synthesis lectures on engineering, science, and technology. Springer Cham, 978-3-031-44033-5. https://link.springer.com/book/9783031440328
21. Tokgöz, E. (2025, February). Artificial bee colony optimization techniques' utilization for intrusion detection systems' analysis. In *4th IEEE international conference on AI in cybersecurity (ICAIC) proceedings*. https://ieeexplore.ieee.org/stamp/stamp.jsp?tp=&arnumber=10848880
22. Morgan, J., & Brenig-Jones, M. (2012). *Lean Six Sigma for dummies* (2nd ed.). Wiley.
23. SPSS Modeler by IBM. https://www.ibm.com/products/spss-modeler
24. SAS by IBM. https://support.sas.com/documentation/onlinedoc/sasc/doc700/html/changes/zgenid-5.htm
25. SAS by IBM. https://support.sas.com/rnd/itech/doc9/dev_guide/dist-obj/javaclnt/javaprog/index.html
26. Palisade. https://www.palisade.com/
27. Roth, T. *qualityTools: statistical methods for quality science*.
28. SigmaXL. https://www.sigmaxl.com/SigmaXL_Features.shtml
29. Precision Tree add-in, Palisade. https://help.palisade.com/v8_7/en/PrecisionTree/About.htm
30. https://help.palisade.com/v8_7/en/TopRank/About.htm
31. SigmaXL. https://www.sigmaxl.com
32. Triptych and Apogee software, Statistical Design Institute. Accessed on June 4th, 2025. https://stat-design.com/Software/SoftwareOverview.html
33. Statistical Design Institute. Accessed on June 4th, 2025. https://stat-design.com/
34. NCSS Statistical Software. Accessed on June 4th, 2025. https://www.ncss.com/software/ncss/procedures/
35. Statgraphics Sigma express. Accessed on June 4th, 2025. https://www.statgraphics.com/statgraphics-sigma-express
36. Six Sigma and Lean Six Sigma, Statgraphics 19. Accessed on June 4th, 2025. https://www.statgraphics.com/six-sigma-lean
37. JMP Statistical Software. Accessed on June 4th, 2025. https://www.jmp.com/en_us/home.html
38. JMP Statistical Software. Accessed on June 4th, 2025. https://community.jmp.com/t5/JMP-Add-Ins/Six-Sigma-Menu/ta-p/22664#

7

Improvement of the Systems

The improvement ideas developed up to this point due to understanding the system better in addition to the improvement ideas that existed prior to starting the Six Sigma project may be very different; therefore, the phases Define, Measure, and Analyze prepare a Six Sigma team for much better improvement opportunities. In addition, improvement of a process requires a conceptual understanding of all the information collected during the prior three phases. Even though Analyze phase may appear as having a system analyzed through analysis of observations for can be improved, Six Sigma team member need to remember all accomplishments in all phases and outcomes throughout the project steps. Improve, the fourth phase of the DMAIC approach requires several tasks to be accomplished by answering several questions:

- What types of waste were observed for elimination and what can be done to eliminate them? One effective way of determining this is to look at or structure a cause-effect diagram that can help to view the big picture for possible solutions.
- Using the outcomes of DMA steps, do the improvement strategies determined make sense within the system? Is the system compatible with the improvement idea applications?
- What techniques can be used for the improvement of the system through the reduction of waste?
- Have the stakeholders been consulted about the possible effectiveness of improvement strategies?
- What are the costs that arise due to implementation of the improvements or changes within the system and are the stakeholder comfortable with them? Cost here doesn't necessarily mean money, it could be time, effort, computational effort etc.

- Is it possible to test the improvement ideas on the system (or a part of the system) for waste reduction? (Assuming the stakeholder is comfortable with the application of the testing).
- Upon testing the system, have there been any updates needed to be applied on the idea?
- What measurable improvement outcomes are attained based on the structured cause-effect diagram designed for improvement ideas?
- Is it possible to apply two or more improvement methods on the system to test the system? This may depend on the costs (e.g. limitation of access time to a network by the users, time, money, energy, system's efficiency related considerations) associated with running the system.

Six Sigma team members are expected to implement improvements to the process after testing them on a set of instances. If the improvements work at an expected level (which is what the customer really wants), then it is important to test the improvements on the entire measured data, process, product, or service. Six Sigma team members should always keep in mind the main objective of the Lean Six Sigma project: customer satisfaction. The following are eight points that are highlighted for a successful cyber security project application that a six-sigma team should also pay attention during the completion of a six-sigma project [1]:

- Assign security responsibilities and roles.
- Identify and document asset vulnerabilities.
- Determine organizational risk tolerance.
- Manage end-user identities and access credentials.
- Manage privileged users.
- Create a culture of security awareness.
- Maintain an incident response and business continuity plan.
- Monitor computing environment to detect potential cybersecurity events.

These bullet points will be revisited throughout this chapter.

7.1 Introduction

The process, product, or service under consideration for improvement has measurements. The analysis of this data tells you a story that you need to first listen to and understand from different perspectives for the defined problem; communication is one way traffic during DMA in which case you are listening to the system about what to be done for improvement. During DMA, you should have a thorough understanding of the entire process, product, or service where the improvement opportunities should be identified, and the variational nature of the data should be clear. In this chapter we will cover several techniques that will help to improve the process from different perspectives:

- Design, duration, insight (i.e., conceptual coverage), and purpose aspects of developed software improvement for cyber security applications.
- Mistake prevention.
- 5S.
- Design, duration, insight (i.e., conceptual coverage), and purpose improvement of networks that relate to cybersecurity.
- Data storage and staff/employee utilization as a part of resource management.
- Employee training/improvement.
- Security assessment from data and risk perspectives.
- General infrastructure improvement.
- Metric improvement.
- Reliability improvement.
- Simulation.
- Data Analytics, Machine Learning, and Artificial Intelligence.
- Time Improvement.
- Improved?
- Improvement worked?
- Value Stream Mapping (Improvement Phase).

It is always important to remember that improvements should be continuous: The improvements should last in the process for a long time (if possible, forever unless better improvements are possible) and make the improvement stable in the process. The improved process should not change, and it should be possible to apply other improvements to the improved process.

7.2 Software Improvement

Software may be developed by a cybersecurity software developer, could be custom made for the company developed by another company, and could be commercial. The improvement of the software may depend on several factors such as its design, duration of service, insight, and purpose. The Analyze section suggests the need for such improvement as a part of the cause and effects realized.

7.2.1 Design

The design of the software may have the strongest impact on the cybersecurity application. The design aspect of the solution includes but is not limited to the following:

- Efficient coding of the programmer.
- Algorithm used for development.
- Software used for development.

The bugs existing in the design of a software solution (which shouldn't be happening) is an ordinary issue that even largest corporations face today. Even though these software solutions go through many testing phases, there may still be unknown reasons for occurrences of the issues due to the complex nature of the software (or tools used for developing the software solution). It is good practice to break down the solutions into several sub-solutions to work out the details of the problems. Six sigma projects can particularly serve well in such situations.

7.2.2 Duration

The speed of software to respond to a thread or its access to a system to secure the system are durational considerations of the software. The speed of the software can be directly related to the speed of the algorithm that can depend on its computational complexity. The impact of duration will be evident from the analyzed data collected in measurements. A practical solution to overcome the duration of responding to attacks can be to simulate cyberattack scenarios to test a variety of commercial scenarios. If the software is custom made, then it needs to be tested for improvement by the developing company. It also makes sense to purchase another software package that will serve for the weaknesses observed in the Analyze phase as a quick reaction to cover the observed gap. These are a few examples that we can provide; however, as one can imagine, there are plenty of solutions that can be designed. Several solutions' cost–benefit analysis can be compared to determining the potential contribution of each solution in a six-sigma project.

7.2.3 Insight (Conceptual Coverage)

A software solution can be as good as its coverage. Let us suppose a company wants to evaluate its cyber security software's effectiveness in a six-sigma project. The existing software packages can be compared with the other similar choices available by other companies. After collecting internal software package data and comparing it with similar external software packages, the results can indicate the shortcomings of the software used for conceptual coverage of the existing cyber-attacks. The improvement in this case can be increasing the conceptual coverage of the software by either purchasing another package that makes up for the missing conceptual pieces for responding to threads or replacing the existing software with the one with broader coverage with stronger reactions to attacks. All different aspects need to be considered when such decisions are made (such as speed, design etc.)

7.2.4 Purpose

The purpose of software developed for cybersecurity applications depends on the end goal of the program which can be driven by priority to respond while insight aspect of software is coverage of the software. Software may have extensive coverage (i.e., insight) however may not be strong in responding to threads due to its overall design. The purpose of software may have a hierarchy of its strength in such applications. For instance, software might have the highest ability to respond to phishing attacks while not so good at responding Drive-by attacks. A six-sigma project may indicate that the company is facing both attacks therefore the weakness (or vulnerability) appears to be responding to the Drive-by attacks. This software might be effective at a low level for all types of attacks except Phishing attacks as well. An obvious improvement would be either purchasing another software that can close the gap of vulnerability gap or replacing the existing one with the one that has stronger reaction to the two types of attacks. One needs to keep in mind that the other attacks should also not be forgotten during the design of the solution in order to have comprehensive coverage of attacks at an optimized level of coverage.

7.3   5S (Five S)

Five S is a Lean technique to improve the existing work conditions to have an ordered, sorted, and clean environment that meets the business regulations. The five "S" cycle components are as follows:

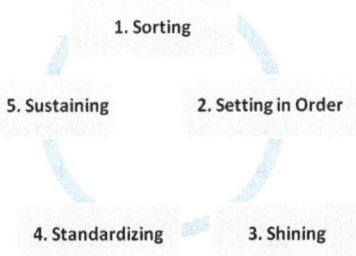

In this cycle of improvement opportunities, 5S elements are as follows:

1. **Sort**. A threat response system is usually a dynamic environment in which response decisions and risks can change anytime during a response to a threat, service activity, or process completion. Too much variation from the expected value (i.e., too much fluctuation around the average) is not welcomed or favored in systems. Sort is one of the methods of overcoming this issue. A simple example in cybersecurity is the sorting of threat handling and time to close incidents, therefore threats to be handled can be sorted in a way to improve operations and they might cause time waste if there is no specific order. A policy needs to be structured for handling such issues based on risk assessment and analysis for further improvement. This massy environment can result in wasting time finding things that we need. Therefore, it is best to have a policy based on severity of risks to be able to improve time to close incidents and threat response for not wasting time to decide the order of incidents to response. The threat sorting process requires continuous improvement as the risks assigned for threats and overall analysis of the operations that may change at any given time. In manufacturing, it is important to sort items that are used often. The items used in the workplace are usually well known, therefore the following can be done by the Lean Six Sigma team for sorting:
   a. Make a priority list of items used in the workplace.
   b. Determine the frequencies of the usage of these items in the list (how often they are needed).
   c. Make an item list including the corresponding frequencies to point out where these items are used in the workplace and how they should be sorted.

2. **Set in Order**. After sorting the threat response system priorities, it is crucial to find the right order to respond to them within the system. This placement is very important and should respect the policy made unless there is an extreme case's occurrence. As many times as the system enters threats that require quick responses, as fast as the responses to these threats should be activated. After determining the order of the threats to be treated, it is necessary to follow the regulations and appropriateness of the threat responses detected. Setting in order in a cybersecurity environment can be done by doing the following on the threat responses:
   a. Design a map of what is expected from the respondent during the operation before changing places of threats for treatment. This is important since the person who is operating the response system should be comfortable with any changes.
   b. Design a new map with the necessary changes by considering what the operator is comfortable with. This can be an automated system with updates reflected automatically on a dashboard setting with what is worked on and what is expected to be worked on in order. Color coding can help with the order of threat detection and completion.

    c. Sketch a map of the threat response mechanism used in the workplace: This map can be used for continuous improvement because the same design can be implemented in different places of the process and used for training newly hired staff.
    d. Consult with the operator to see if the map is realistic for implementation (if the operator doesn't design the map.) This map can be a dynamic one to track live changes in the operations and respond to threats. Such an approach can reduce many NVA activities...
    e. Test whether the new "threat response map" is working well when it is used.
    f. Make the necessary changes to the map after implementation.
    g. Finalize the designed map.
3. **Shining**. Keeping the system components and the system "shining" to have the best productivity out of the system is also critical. For instance, during threat response activities, keeping threat responses free from NVA activities is particularly important for them to operate smoothly. Updating the tools used for responses is a part of shine to make sure that the threats can be responded at the best possible speed and method which can decrease the down time of threat responses and speed up the time to response incidents which may increase production. Such an approach is expected to result in an increased percentage (or rate) of incidents responded and resolved while the associated costs may increase (if the customer is comfortable with it.) Lean Six Sigma team can set up cleaning regulations by doing the following:
    a. Continuously revise the response capability of tools used in the inventory for changing conditions and how often these items are used at the workplace. Clearly communicate the results with the associated users and publish a guide/manual as needed.
    b. Make a chart of how often, when, and how the inventory used should be updated.
    c. Place/share this chart in the workplace at a location that is easy to see.
4. **Standardize**. Continuing the implementation of the first 3S steps will yield to an organized work environment that yields to a standardized process. Standardization of the changes is the responsibility of the person operating it. Automated systems can be helpful in standardizing certain aspects of the 3S while human involvement and decision making would always be a part of Standardize to work out operational changes in decisions and customer's decisions.
5. **Sustain**. Make sure that the first 4S components are implemented continuously on the system. As pointed out for Standardize, automated systems can be helpful in sustainability of some aspects of the cybersecurity operations while other operations would require decision makers. For instance, one can sustain a process by placing a map of threat reaction operations and changes on a chart of cleaning threats in the system for ease of access. Following the policies can help with the sustainability of the operations.

7.4 Network

Network improvement for cybersecurity applications requires a good understanding of the network elements. The design of the network for the cybersecurity solution to take place is critical for foundation, duration, transportation, insight, and purpose of the solution. The network elements usually have a complicated system design in which the process steps require an optimized level of cybersecurity application. Optimization here can mean maximizing and minimizing variational elements of the network. For instance, the goal can be maximizing the security agents access to the system when the threads are detected to be increasing in parallel to minimizing the number of servers employed for serving the system so that the vulnerability to the system access can be minimized. This design can be dynamic and changes over time. The complexity of cybersecurity solutions depends on their design.

The current design of the system may have flaws and understanding of these flaws would depend on the Measure and Analyze sections. The network design might be causing more cybersecurity problems than it intends to. For instance, too much network access given to too many employees with the overcrowded network design can cause chaos within the organization. The clearance provided to too many personnel to override the same system is a simple example of a poor network design that can impact the security level at any given time as the experience suggests. Analyzed measurements would point to improving the network design in a way that it can serve at an optimal level. The following are additional considerations on the network operations for improvement opportunities:

- Network access.
- Network size and operational volume.
- Network elements.
- Network coverage.
- Network connections and flow.
- Network's duration of operation.
- Cloud versus local storage and clearance for access decisions.
- Levels of clearance to access network and systems.
- Network costs.
- Network design and changes needed in the design.
- Network authentication decisions.
- Training of staff for network access and decision making.

7.5 Improving Design

We previously pointed out network and software design improvement. In this section we will focus on some other aspects of design that relate to cybersecurity applications. These aspects include

1. Cyber solution design.
2. Alternative designs.
3. Transportation design
4. Documentation design.

We will focus on VA, BVA, and elimination of NVA activities for applications. In this section, we'll cover how to improve these designs during the Six Sigma project.

7.5.1 Cyber Solution Design

The cyber security solution needs to be designed in such a way that the staff can be comfortable to the highest degree, otherwise it may cause unproductive results on day-to-day operations. A Six Sigma project's solution would be ideal if it is possible to design a solution by which the staff is comfortable while working and has the best productivity during the work hours. This may seem to be a light improvement consideration during a six-sigma project by cyber security professionals, however its importance cannot be disregarded noting that human involvement in the digital world is the most crucial component that can cause most of the mistakes. Unhappy employees in the workplace might not cause more problems than expected because of undesired solutions produced. One way to test this aspect would be creating several six sigma solutions and surveying employees to determine the best option favored by many end users.

This improvement of design is possible by observing the activities with the software used for a short period of time (if necessary, for future reviews). This improvement is particularly useful for reducing the software that is not used in applications that can increase the productivity of the staff. The following steps can be followed for design improvement at the workplace:

- Collect data on access of employees to certain software under consideration.
- Design a survey that would allow you to understand employees' view of the use of the system components in place.
- Analyze the survey results along with the functionality of the software. If a software package use is a "must" in the system, then you may need to clearly state its importance of use and collect data on opinions of simplifying the use of the software. The survey design must be done smartly.

- The possibility of the new design should also be tested with the employees.
- If the staff is comfortable with the new design, implement it by pointing out the new locations of the items with colored tapes etc. This is particularly useful for the staff not to forget which item should be where in case of relocation of items.
- Post the designed changes on a board.

7.5.2 Transportation Design

The design of the transportation of entities can impact cyber security operations and it can result in issues with access level changes and responses to threats. There are different forms of transportation including but not limited to stored tools for reacting to threats, clearance to access a variety of systems etc. Based on the ever-changing requirements of transportation, the following steps for improvement can be followed:

- Find out/observe the tools and techniques used for transportation and question their effectiveness; question whether it is the best way to do the transportation.
- Determine the route of transportation and sketch a map of it.
- Determine the best possible optimal way of transportation, the necessary tools for implementing it, and determine the necessary rerouting of the pathways for making improved transportation possible.
- Depending on the type of transportation, either sketch a map of the redesigned routing or place trace marks on the transportation map pointing out the best possible way to be followed from one location to the other.

Communication is one of the major issues in the workplace that can harm the cybersecurity operational decision and miscommunication occurring around the levels of access clearance changes can be one of these issues. Phishing attackers can take advantage of such cases.

7.5.3 Document Design

The steps taken to design the solution need to be carefully documented in addition to the entities of the system used before and after improvements. One challenge that appears to be in cyber security documentation is the overcrowded nature of the explanations. One may need to divide-and-conquer the documents with more visuals used (what is so called One-Point Lesson).

Another aspect of documentation is identifying and documenting asset vulnerabilities. The identification of asset vulnerabilities itself can be a six-sigma project which can be a lengthy process with extensive data collection and analysis required.

7.5 Improving Design

There are many documents used in workplaces that can be improved such as documents used for decision approvals requiring signatures, explaining work procedures etc. These documents may not have the best structure that the customer is willing to pay for and they can be improved. One way of improving these documents is as follows:

- Study the current structure of the document and determine whether it has the best structure that is needed by the business.
- Determine the weaknesses of the documentation by observing what appears unnecessary as a part of the process that the document is designed for. Also determine whether the document has summarized information if there is any.
- Eliminate or take out everything that is unnecessary to appear on the document.
- Redesign the improved document for the approval of the project owner.

7.5.4 Duration

The duration of documentation is the time it may take to go over the document. In a cyber-attack which requires quick reaction may require to go over a set of documents however if these documents are overcomplicated then the reaction will be delayed which increases the tack time on the response. It is possible to improve documentation with more visuals or "coded" language that may be understood by the cyber security professionals.

7.5.5 Infrastructure

The organization of the document needs to be in such a way that the results conveyed are easy to access and followed by the employees. Overcomplicating the language in the document would not help any other employee, therefore an improvement opportunity can be on the infrastructure improvement of the documentation method.

7.5.6 Insight (Conceptual Coverage)

The conceptual coverage of the documents can be designed in such a way that priority can be given to the most common issues that occur. This is a "weighted" approach in a way that the count of the maximum attacks detected can be attained upon the completion of the measure phase in which case the corresponding response document can be the one on the top of the list. This can reduce the tack time for ease of access. One can apply a "tree" approach in which case the main branches of the tree consist of the main categories of thread responses while the subcategories consist of the smaller branches. This branching methodology can be as deep as it can be, although one needs to be careful about the takt time of reaching the response method for an incident.

7.6 Employee Training/Improvement

Human error is one of the top reasons for mistakes occurring in cyber security applications like in everything else. For instance, as a result of 601 individuals surveyed in companies that have a data protection and privacy training (DPPT) program and who are knowledgeable about the program reported the results. Managing Insider Risk Through Training and Culture Report also revealed 66% of the data protection and privacy training professionals that were surveyed labeled their employees as the "weakest link" when attempting to safeguard their organization from cyber threats. The most common human mistakes in cybersecurity are

- clicking on links.
- opening unknown attachments.
- entering personal or confidential information into a seemingly friendly (and familiar) account.

A technique called social engineering is utilized by hackers to take advantage of human behavior to successfully pull off a ruse or scam. There are critical key points that can cause a mistake because of human error:

- Lack of experience.
- Lack of standardization.
- Lack of training.
- Forgetfulness.
- Misunderstanding.
- Bad identification.

The four of the eight cyber security improvements needed to reach the next level of maturity can be the following:

- Assign security responsibilities and roles.
- Manage end-user identities and access credentials.
- Manage privileged users.
- Create a culture of security awareness.

Such considerations are all depending on employees. This is a strong indicator of the need for employee training for improvement. This training can be leveled based on the credentials of the employees and their clearances to access subsystems. Employee improvement is mainly possible by focusing on the training of the following:

- Emphasizing employees' responsibilities to protect the company data.
- Awareness of document management and notification procedure uses.
- Right password choices and careful use of the passwords.
- The importance of not using unauthorized software.
- The challenges that can be faced are because of using the internet for purposes that are unrelated to work.
- Checking or responding to emails that are irrelevant to work or from unknown resources.
- Recognition of common cybercrime and information security risks, including social engineering, online fraud, phishing, and web-browsing risks.
- Social Media Policy.
- Mobile device use and its consequences on cyber attacks.
- Protection of computer resources.

7.7 Poka Yoke (Mistake Proofing)

Poka Yoke (which is also known as mistake proofing) is a technique with the zero-defect goal that aims to use devices and/or procedures effectively for error detection. The strategy relies on the fact that an error made increases its economic impact over the time it occurred and the time it is realized. In cybersecurity the time difference between an incident occurred and the time it is realized. An important consideration in this approach is to determine when and how to react. The design choice in Poke Yoke is called Poke Yoke design while the detection mechanism on the process is called Poka Yoke process. An example would be catching an incident to occur before it occurs.

7.8 Simulation

Simulation is a strong tool that can be used for generating scenarios and testing an improvement technique on the generated scenario. There are several simulation tools that can be utilized in applications; however, the simulation can also be custom made depending on the problem tackled. The classic simulation steps after defining, measuring, and analyzing the problem related instances include

1. Model Translation.
2. Model Verification.
3. Model Validation.
4. Experimentation.
5. Analysis.

Model translation is the step in which we design a simulation model with the corresponding statistical distributions assigned to the corresponding instances. Model verification is to check how much the developed simulation model matches with reality. In the case the model doesn't match well with the collected data (to a certain statistical percentage) then it needs to be validated. The experimentation step can be testing the model with experiments with what follows as the analysis of the experiments [3]. If we consider cybersecurity applications, resistance to vulnerability can be simulated and tested by following the corresponding steps:

- Use your measured data to derive statistical distributions that will demonstrate your existing system in the simulation solution. You need to make sure that the distribution is a good representation of the system in place with its vulnerabilities and strengths.
- Design a solution that potentially eliminates the weaknesses of the existing system as a distribution.
- Run your simulation on the existing system to test its effectiveness over a set of instances over time to determine if the solution can be effective on improving existing system vulnerabilities.
- Change instances on your simulation model to see if a better design can be used for declining attacks.
- Choose a simulation model that fits well with the desired results and use its parameters to develop the corresponding solution.

The data collected in the Measure phase plays a critical role in the development of simulation solutions. Statistical calculations such as average and standard deviation along with the statistical distribution to be a good representative of the data are critical in success of the simulation solution. A benefit of simulation is its ability to give you measurable outcomes for comparison with the existing system capabilities. These measurable outcomes are approximate in nature noting that the statistical distribution that we fit to the data approximates the data, and the data can be time dependent therefore the seasonality of the data can be impactful on your simulation results; these are critical considerations if there is interest in implementing a simulation solution. Simulation can be particularly useful in cases when the system does not have much variation (regardless of seasonality) in addition to having a good statistical distribution fitting to its data. Therefore, simulation can be a good tool to use if the nature of the exiting data and system is low in variation and has a good mathematical representation that reflects its nature in a timeless fashion.

For instance, in the case of a new version of Dridex release, with new indicators of compromise (IoCs), how can one test if the existing security controls can detect it? In addition, newer techniques may have been added to the latest strain with newer system exploitation methods, which may have made yesterday's methods absolute. Simulation of the underlying attack methods become critical due to the need of preventive controls to catch a specific strain. In this case detective controls can determine if any strain that exhibits the same underlying attack method or technique can be caught. We need to note that detection controls rely on tools such as machine learning and AI that are behavior-based to identify malicious activities while static IoCs are used as preventive controls.

An example of a cybersecurity simulation application is breach and attack simulation in which case automated testing tools can be used for filling in the gaps for point-in-time assessment.

7.9 Data Analytics, Machine Learning, and Artificial Intelligence

Extensive applications of data analytics in cybersecurity are not surprising given the data driven nature of the field. There are books written in this area of interest and it is no brainer that the improvement of your problem might be directly associated with one or more of the techniques used in this field. The classic data analytics concepts are built on data preprocessing, visualization, correlation, regression, forecasting, classification, and clustering. Some of the most recent advanced techniques require machine learning applications in which case collected data sets can be simplified by using algorithms to determine the most relevant features that are associated with the expected results. Therefore, unlike in addition to the classical techniques that required the use of statistics, machine learning is the use and development of computer systems that can learn and adapt without following explicit instructions, by using algorithms and statistical models to analyze and draw inferences from patterns in data. Data analytics and machine learning work hand in hand. An important remark we need to make is the young age of both data analytics and cyber security fields; while there are known applications of data analytics in cybersecurity, there are many others that have not been applied in this area yet. These applications and their explanations would be beyond this book; however, we will give several examples to such applications to increase the curiosity of the interested readers:

- Data analytics is applied to support IT security experts by identifying and improving the novel or obfuscated attacks in a company regardless of their origin being inside or outside the company.
- Combination of visual analytics along with machine learning techniques in order to create a human–machine approach for reducing false positives flagged by a system for insider threat detection.
- Dynamic opcode analysis by using a machine learning technique on a data set (of 610–100 K features and depth of 48 K samples) with the improvement indicating that dynamic opcode analysis can detect malware from benign ware with a 99.01% accuracy rate, using a sequence of 32 K opcodes and 50 features.

As can be seen from these examples, cybersecurity goes beyond the typical data analytics methods used on ordinary data sets and integrates some of the state-of-the-art techniques for further improvement of the field. Artificial intelligence (AI), with machine learning being one of its applications, is the science and engineering that attempts to make machines behave as close as possible to humans [14]. Artificial intelligence can efficiently analyze user behaviors, deduce a pattern, and identify all sorts of abnormalities or irregularities in the network. It's much easier to identify cyber vulnerabilities quickly [13].

Detection of abnormal activities through artificial intelligence in a cyber environment requires removal of noise or unwanted data to enable security experts to understand and improve the environment better [12].

A simple example of an advanced six sigma project can be the third bullet point listed above:

- Define: How can the detection of malware from benign ware be improved?
- Measure: Collect the features and opcodes that relate to the dynamic opcode performance.
- Analyze: Reduce the features and opcodes that relate the dynamic opcode through cause-effect analysis.
- Improve: Design a machine learning algorithm with high level of accuracy to detect malware from benign ware.
- Control: Design a control chart that helps to detect the out-of-control points in malware detection. Control if the algorithm can be sustainable in applications to other data sets.

The control section is a critical component of this application noting that the machine learning techniques can be limited to the data tested and may not be applicable in other data settings. The sustainability performance of machine learning needs to be tracked over time as a part of continuous improvement. Several AI techniques, including machine learning, and data analytics techniques meet in the driverless automated cars maneuver and object detection.

7.10 Time-Driven Improvements

Improvement of time-related activities in a cybersecurity Six Sigma project may be very important as timely response to systems' cybersecurity management is necessary. Lead Time is one of such a time-driven concept, and its improvement for the general process is an important aspect of overall improvement. Identification of Lead Time is typically transparent on Value Stream Mapping with all Six Sigma project related operations are clearly stated. On such a map, all primary activities that include VA, NVA, and BVA activities are identified with the corresponding times spent for them. It is well known

7.10 Time-Driven Improvements

that the speed of reaction to an attack needs to be quick with possibly a Kaizen event; however, looking at a bigger picture may indicate the need for improvement in this quick reaction over a longer period by employing a six-sigma project. Some of the aspects of time improvement include the following:

- **System's Downtimes**: The down time is the time frame that a machine is unexpectedly down for a while. This time frame usually does not add value to the process and the only way to improve machine downtime is by taking immediate action to fix it. If downtime is a must in the system, then it should be considered as a part of the improvement strategy.
- **System's Setup Time**: It is the time that it takes to set up a machine after the completion of a successful task from one unit to the beginning of processing of the next item. This also related to the "Single Minute Exchange of Die" concept in six-sigma applications.
- **System's Performance Time**: Measures performance period of a machine during a production or a process. A machine may not have the best timing/performance expected from it if it is not clean, old, broken etc. based on its credentials such as hardware age and software versions used in this machine which are also impactful on performance. The way to improve this timing is by maintaining and controlling its performance periodically with the changes to these machines implemented as needed.
- **Production/Process/Service Timing**: Production, Process, and service timing is the amount of expected time to produce an item or complete a process or service. The best way to improve this timing is by respecting the takt time of the production/process/service.
- **Staff and Time Management**: Improvement of staff and time management depends on the skills, tasks, and completed works that include the following:
  - The right task assignment to right people who are skilled to fulfill the associated expectations. If such workforce does not exist, upskilling the employee talent may be needed, and it is particularly essential for upskilling employees based on developing software patches and tasks.
  - Do the employees know the system well enough to be able to work on the system with fully expected capabilities?
  - Right cybersecurity and systematic application design to save staff time for finding items, machines, and structuring effective communication at the workplace. Communication is the key!
  - Written cybersecurity information can be very crowded and overwhelming to follow for some of the employees; Is there right amount of information on the documents that is minimized for best timing performance expected from the staff. Designs that translate such overwhelming paperwork into flowcharts can be particularly helpful for reducing the overwhelming paperwork.

- **Inventory-dependent Time Management**: Inventory of a system may include any entities such as hardware, software, database, and other physical entities, and time management of accessing or using such inventory is critical for successful operations within the system. If the inventory entities that are necessary for producing the output of a project are not in the inventory or not functioning properly, then the production either may not occur, or the system's design would not be functioning at full capacity due to missing entities. Such issues may cause additional waste such as machine performance time, staff time, production timing etc. due to not achieving the best performance from the system. Important aspects of inventory time management and improvement methods include but not limit to the following:
  - Structure a list of inventory entities with possibly specs of the entities such as year of purchase or subscription, update times, abilities etc.
  - Identify the expected quantities of entities needed in the inventory and the locations of these items needed for producing output for the project.
  - Designing smaller inventory locations that are called *Local Markets* for storing and using these frequently used items for ease of access; *Local Markets* can be designed for both hardware and software; An example of a Local Market can be a cabinet placed in the workplace. This application can be particularly useful for having the inventory close to the physical areas of application or software access points for ease of speeding up processes. Storage of actively used units closer to the areas of application is one practical improvement for a Six Sigma project in addition to installing software in servers that specific machines can access easily compared to other servers.
  - Design of push and pull systems can be designed for either pushing from the inventory or pulling by the system from the inventory to improve the duration of utilization of the inventory:
- A push system that is built into a system would require the inventory to push entities as needed to the location of usage or local market. In this case the downstream operation is pushed items from the upstream (i.e. inventory).
- A pull system requires down stream's pulling entities from upstream and in the case of inventory, system operations or operators that need entities from the inventory can pull items as needed. Pull logic can also be applied to replenishment of the Local Market.
  - Local Markets need to be kept up to date as otherwise there will be a shortage of entities to be used within the region.

7.11 Do We Have Effective Improvements?

Questioning whether the improvements worked or not is a necessity without a doubt. The expectations in the case of a Six Sigma project can be quantitative difference between the pre- and post-improvement stages of the Six Sigma project period. Hence, questioning whether we have made improvements, and they worked or not should be one of the first

7.11 Do We Have Effective Improvements?

questions that need to be asked by a Six Sigma team at each phase during the improvement. The Six Sigma team can seek answers to the following questions for effectiveness of improvement ideas:

- Were the improvements effective at every single step of the process to be improved from top to bottom?
- Was there a good amount of data collected to have a significant evaluation of the system considered for the Six Sigma project?
- Was the analysis implemented correctly?
- Were the VA, NVA and BVA activities of the process identified, evaluated, and improved effectively?
- Was all the waste categorized and worked on during the project?
- Were the quantifications of Six Sigma improvements contained strong theoretical results after eliminating NVA activities?
- Were the employees' (or staff members') expectations incorporated into the improvement plan?
- Did the Six Sigma team document the improvement ideas?
- Were the budget expectations met for completion of the project?
- Was the project completed on time?
- Were all the steps of the process revisited to determine waste?
- Is it possible to continuously improve the newly designed system? That is, do we have a system that is open to improvements that would be long lasting and not returning to its prior stages?
- Were the stakeholder's requirements and expectations met?
- What savings are expected upon the implementation of the improvements?
- Which wastes are expected to be eliminated and what benefits are expected from such elimination?
- Were the stakeholder's quality expectations met based on the improvement ideas developed?

Brainstorming within the team needs to be professional and respectful and should not affect the project's progress. Upon asking and answering questions by the Six Sigma team members, a detailed report of the project that is concise and easy to follow should be prepared to explain the details of the work completed for the project, the improvements applied and the difference between the system that is based on pre- and post-Six Sigma project work. The project results need to be presented to the stakeholders, and the Six Sigma team should design signs and guidelines for the system's personnel for the implementation of the improvements within the system.

7.12 SMED (Single Minute Exchange of Die)

Setting up entities and changing over from one operation to another may have NVA activities that can be eliminated. Such changes can be eliminated by using SMED (Single Minute Exchange of Die) technique. Changeover time is the period between the production of the last piece of the previous production to the first good piece of the next production within the system. Changeover time may occur in any place where production occurs such as a process, a machine, a server, a network, etc. The application of SMED focuses on the waste that may occur during the production or processing by observing the system that operates on it. Regardless of whether it is automated or not, reduction of realized changeover NVA times add value to the process since it increases the effectiveness of the system. The following are some of the observations that are made by applying SMED based on NVA identified while operating a system:

- **Manpower**. There may be numerous operators operating the same machine within the system altogether. Lack of teamwork may result in changeover time NVA.
- **System Parts**. A system part's design or choice may impact change over time and therefore the system would not perform at its best. Selecting the right system entities with the right features and properties may be impacting the SMED timing.
- **System Operations**. The design of the operations may be causing the change over time to produce NVA and rearranging such operations may change the overall performance of all systematic entities.
- **Production (Batch) Size**. The production of the expected results during each subprocess based on the size/quantity of the expected output may have an impact on the time spent by the next subprocess to produce the expected output. If the downstream operation would not produce enough amount of output expected, then the upstream operation that receives the output from the downstream would not have any work therefore this would cause NVA. The production should be at an expected level for the system to continuously produce the results at expected levels.

Operations can be defined as internal and external as follows:

- **Internal Operations**: These operations are defined to be operations that are internally related to the system during the work's completion.
- **External Operations**: These operations are defined as the operations that can be completed outside of the system while the system is still operating.

Activities that are internal and external to a system continue to take place during the changeover time. During the focus of a system as a part of a Six Sigma project, external activities that occur within the system can be moved out of the system for improvement purposes during SMED since these activities can be fulfilled independently from the system. Internal activities cannot be reorganized or moved out of the system's SMED; therefore, removing the external activities outside the changeover time can be a VA to the process while internal activities cannot be removed from the SMED considerations therefore they are considered as BVA.

7.13 Did the Improvements Work?

We questioned the effectiveness of the improvements and now it is time to check whether the system has the reflection of the improvements or not. Hence, after questioning the systematic process improvement in the way the stakeholder and the Six Sigma team members expected, the next step is to determine whether the improvements worked or not. This requires several steps that need to be taken to determine the improvement within the system:

- **Pilot Region**. A pilot region is a smaller portion of the larger system or process that the Six Sigma project can be tested for its effectiveness. The selection of the pilot region depends on the scope of the project. After observing pilot region outcomes, the improvements can be implemented in the larger system by using the same technique. The Six Sigma team needs to make sure that the pilot region is selected carefully to ensure that it is appropriate for the system's functioning.
- **Pilot Period**. Just like the pilot region, a pilot period for measuring the success of the Six Sigma solutions can be selected. This idea allows the Six Sigma team to implement the improvements within the system (or pilot region of the system) for a certain period to scale the reaction of the system to the expected improvements. The Six Sigma team needs to make sure that the period is selected carefully to ensure that it represents the system's general timeframe.
- **Duplication**. If the system to be improved is a small portion of a bigger system that is not possible to implement the improvements during the operational hours, then it is best to create a smaller working environment where the process is duplicated for testing the necessary changes. An example of this is the creation of digital twin that serves as the system's duplicate in a database.
- **Full Implementation**. A full implementation of the Six Sigma project can be conducted on the entire system without any pilot region or period that the necessary changes right away if there are clear indicators that the improvements would work within the system without a doubt.

The final step here is the documentation of the above-mentioned findings. Such documents are used during continuous improvement efforts as a part of "lessons learned" as well as "best practices" and "worst practices". There are several questions that need to be answered, and comments addressed by the team that are raised by the stakeholders and such questions should also be incorporated into such reports. Measurements, pilot region(s), pilot period(s), and the reasons that cause the choices made should be explained by the Six Sigma team within the documentation. The last phase of DMAIC is the Control phase and it requires monitoring the system with more updates and effectiveness of the system based on the measured project related systematic performances.

7.14 Value Stream Mapping—Phase of Improvement Applications

A VSM (Value Stream Mapping) of a process is a detailed map of the process from the beginning with the inclusion of the source (i.e. supplier) to the output delivered to stakeholders by using universal icons. Such a map allows a Six Sigma team to have a compact view of all the entities and operations of the process and system. Such a condense representation eliminates the need for going over pages and pages of information to see the big picture of the operations, entities, and works completed within the system with the associated timelines. The prior two stages of VSM required the Six Sigma team to have a broad understanding of not only the overall operations with their timings but also point out the locations of possible improvements with NVA activities. VSM's improvement phase requires sketching the VSM of the new system after applications of the improvements. The information to be incorporated into the newly designed system is the same as the initial phase of VSM:

- Suppliers of inputs.
- Existing operations (Push and pull inventory etc.)
- Local Markets (if exists).
- Entity movement or transportation of entities within the system.
- Lead Times of deliveries, operations, and process completions.
- Systematic Entity Operations.
- Timing at each step of operations.
- Operational costs.

7.15 Prevention of Mistakes

The duration of unrealized mistakes can have a big impact on cybersecurity related task handling. Reacting to such mistakes prior to their occurrence would eliminate possible waste that may occur. Therefore, prevention of mistakes that may occur due to vulnerabilities and some other aspects of cybersecurity is a need for the work to be completed. Advancement of systems during the initial phase of the Improve phase can be detailed by using a process mapping and identifying waste related occurrence and issues by using 5W and how questioning can help. Prevention of mistakes before they turn into waste can be one of the best possible actions and there are several techniques that we will cover next.

7.15.1 Improve Through Controlling at Each Phase

Many systems have subsystems that need to be checked individually. Controlling such a system at each subsystem level is the application of control at each subsystem level. This is one way of improving the process with a control mechanism placed at each subsystem. Some of the systems require such a control mechanism throughout all subsystems therefore such a decision needs to be made by the Six Sigma team based on the project's scope. Some of the considerations include the following during this type of improvement:

- **Use of a Map**: Using a detailed process map to understand the improvement needed on the existing system can be helpful. Such a process map can be designed during the Measure or Analyze phase of DMAIC (e.g., SIPOC diagram for the entire process.) For instance, there may be failing systems in several areas of a network system during communication within the network and implementing a full control system throughout the network can help with the implementation of the improvements.
- **Systematic Feedback**: Design of a feedback system as a signal of process issues when a mistake occurs at a phase would help with the detection of possible waste, therefore, a control mechanism with such a feedback system can be beneficial.
- **Possibility of Correction**: Identification of correcting a mistake that occurs need to be identified. Reacting fast to such mistakes with the correct measures taken in a proper way is essential when it comes to communication and design of the feedback system. If such a correction is not possible then it is best to identify its consequences.

Improvements by the application of a control mechanism (whether manual, automated, or a mix of the two) as a part of the process have advantages and disadvantages:

Application Advantages
Reduction of waste within a process through strict control at each subprocess requires fast reaction. Contamination of the subsystem that experienced an issue may be a need depending on the system therefore the interacting subsystems with such a subsystem need to be handled quickly. Control at each phase or subsystem can eliminate any possible gaps that may be experienced with surrounding subsystems. Such an action can prevent mistaking a similar again within the system if there is human involvement. Eliminating. Such a quick response to the previous phase in the process has the following advantages:

- Preventing the likelihood of similar mistakes and associated waste.
- Limiting the waste occurring stages of the process would keep the system less busy than usual.
- Realization of the waste of time would help with its possibility of correctability if it is realized early.
- Right level of quality for checking the system for such issues to meet customer requirements.
- The overall process with its subprocesses would be under intensive control for detection of waste.

A combination of controlling at each phase and quick responses in the process can result in extensive amount of waste reduction.

Application Disadvantages
There are several disadvantages of controlling a system at each phase:

- Design of a control mechanism that requires checking the system at each phase can be costly. Such a costly action needs to be discussed with the stakeholder on its implementation. When it comes to cybersecurity related issues, it is the best to apply the best possible measures to the systems therefore it can be the best to implement control at each phase even if it is costly.
- In the case of progressive systems, the pre-phase within systematic processing can be interrupted for the identification and the mistake occurred with a cause of stopping the system often. Such a disruption in the subsystems can cause a delay on the works to be completed and can result in other waste during the production or service.
- If there is any personnel involvement needed for subsystems' control, then finding such personnel and the associated costs need to be incorporated into consideration.

It is best to use controlling system at each phase when mistakes of the system are reduced at the highest level with the best possible outcome attainment from the system.

7.15.2 Improving Through Implementation of Control at Some Phases

Identification of the phases with defects during production or service is needed in some systems. Such a system may have several subsystems that work effectively and are known to have no issues therefore they do not require control while other parts of the system may require attention and therefore implementation of control. Such phases are typically either known or can be identified by collecting data during production or service. There are several advantages and disadvantages of using such an approach during a Six Sigma project:

7.15 Prevention of Mistakes

Application Advantages

- The cost of the adoption of control structures would be limited to the places with issues detected as highly reliable subsystems with no issues would not require such attention.
- The focus would be on specific parts of the system with improvement opportunities focusing on such portions of the system.
- The system interruption due to the needed changes and improvements would be limited to the locations that require attention therefore the least amount of cost for corrective actions would likely to occur.

Application Disadvantages

The focus of the approach relies on the reliability of some of the existing subsystems; therefore, if mistakes occur in the places where reliability is assumed then several challenges may occur:

- A failure in one of the subsystems that are assumed to be reliable can be costly therefore when mistakes occur at some of the unexpected locations, defective and costly production or service may occur.
- In progressive systems, issues may be carried over from one stage of the system to another therefore an issue that occurred in a reliable system may cause issues in other parts of the system and keep the service or production busier than usual without realization of the source that causes it.
- The systems that appeared to be reliable may not be reliable, and this would cause more NVA activities. This may be due to the limited data collected from the system that indicates the subsystems to be reliable while they are not.

It is best to implement control at some phases of the process when there is certainty of reliable subsystems exist while some of the subsystems are unreliable with certain types of mistakes occurring, and the mistakes can be observed by evaluating the collected defect data.

7.15.3 Improvements Without Any Control

Implementation of improvements without any control would be typically possible when the service or production is highly reliable. In such a case, the occurring mistakes during the subprocesses are insignificant, therefore such issues may not require control over possible defects' occurrences. Another consideration in this case is deciding upon cost–benefit analysis: If the consequence of the defect is less costly then changing the system then control can be ignored. Such a system can be improved with a reduction on the likelihood of the mistakes in the system. The overall process with its subprocesses should be evaluated to ensure that it makes sense not to implement any control mechanism over a system. The stakeholders need to decide on such critical matters.

Application Advantage

- There is no cost to the Six Sigma project or the stakeholder for handling mistakes during the service or production if the service or production is highly reliable without defects and it is mainly defect free.

Application Disadvantage

- It can be costly in the case when the high reliability of the process turns out to be unreliable. Such issues can cause systematic failures.

It is best to use this technique when the high reliability of the service or production has been proven, or the system does not need any control due to its infrastructure.

Question. What is the best strategy to decide the control level implementation during a Six Sigma project?

Answer. The stakeholders are at the center of such a decision with a certain level of cost and benefit trade-off occurring between defects and control mechanism establishment for safety. The reliability of the system is the key element of the decision while it is important to ensure that such reliability is not hypothetical but well proven. The consequences of not placing control mechanism with a cause-and-effect diagram can be beneficial by implementing the aftermath of the issues faced. In the case when the service or production is typically defect free, it may be ideal to not designate a controller within subsystems for checking mistakes. The final decision needs to be made by the stakeholder with the consequences of the choice clearly outlined. Reliability of the subsystems needs to be checked carefully by collecting and evaluating data over extensive periods; System logs can help with such analysis. The data evaluation of the overall system needs to reflect the general behavior of the defect production within the system after subsystem analysis; therefore, the comprehensiveness of the data is needed during identification of the systematic reliability.

Chapter 7 Exercises

Exercise 7.1 Find software used for cybersecurity applications and explain its strength. What can be done to improve the design of such software in applications? Please explain.

Exercise 7.2 Find an area of emerging cybersecurity issue for which 5S application appears to be an important need of application. Explain your reasoning.

Exercise 7.3 Find an area cybersecurity application that appears to have design improvement is necessary in application due to the issues faced. Please explain your reasoning. What can be done to improve such a design?

Exercise 7.4 Give an example of a cybersecurity application for which controlling at each phase is essential. Please explain why you think it is important to apply such a control mechanism when compared to the others.

Exercise 7.5 Give an example of a cybersecurity application for which controlling at some phases is essential. Please explain why you think it is important to apply such a control mechanism when compared to the others.

Exercise 7.6 Is it possible to have a cybersecurity application for which improvement without control is possible. Please explain your response.

Bibliography

1. Benz, M. 8 *Cyber security improvements to reach the next level of maturity.* https://www.cioapplications.com/cxoinsights/8-cyber-security-improvements-to-reach-the-next-level-of-maturity-nid-1939.html
2. Penetration Testing. https://www.guidepointsecurity.com/vulnerability-management-and-penetration-testing/
3. Applied Simulation Modeling, Seila, Ceric, Tadikamalla, Thomson—Brooks/Cole, 2003
4. Arcidiacono, G., Calabrese, C., & Yang, K. (2012). *Leading companies to lead processes: Lean Six Sigma, Kaizen Leader and Green Belt Book.* Springer.
5. Cyber Security Training for Employees. https://www.travelers.com/resources/cyber-security/cyber-security-training-for-employees
6. Carrascosa, P., Iván, K., Kumara, H., & Huang, Y. (2017). *Data analytics and decision support for cybersecurity: Trends, methodologies and applications.* Springer.
7. Runkler, T. A. (2012). *Data analytics: Models and algorithms for intelligent data analysis.* Springer.
8. Data Breaches, Experian. https://www.experian.com/data-breach/2016-ponemon-insider-risk?WT.srch=2016_insider_risk_pr
9. Oxford Dictionary. https://www.oed.com/
10. The role of AI in cybersecurity. https://blog.eccouncil.org/the-role-of-ai-in-cybersecurity/
11. Search Control Methods in Deep Blue, Semantic Scholar. https://pdfs.semanticscholar.org/211d/7268093b4dfce8201e8da321201c6cd349ef.pdf
12. Tokgöz, E. (2024). *Quality and lean Six Sigma applications for industrial engineers.* Springer Nature. https://link.springer.com/book/9783031557392
13. Tokgöz, E. (2024). *Quality and lean Six Sigma for engineering technicians, synthesis lectures on engineering, science, and technology.* Springer, 978-3-031-44033-5. https://link.springer.com/book/9783031440328

14. Tokgöz, E. (2025, February). Artificial bee colony optimization techniques' utilization for intrusion detection systems' analysis. In *4th IEEE International conference on AI in cybersecurity (ICAIC) proceedings.* https://ieeexplore.ieee.org/stamp/stamp.jsp?tp=&arnumber=10848880

Control for Sustainability of Improvements 8

Control mechanisms in cybersecurity are essential for adequate operational performance of the cyber security related systems and optimization of these performances [2]. Control as the last phase of DMAIC approach in Six Sigma requires establishing controls over the system for the system to perform in the way for it to operate by maintaining the applied improvements. The right control mechanisms need to be structured in a way to enable the best operational performance.

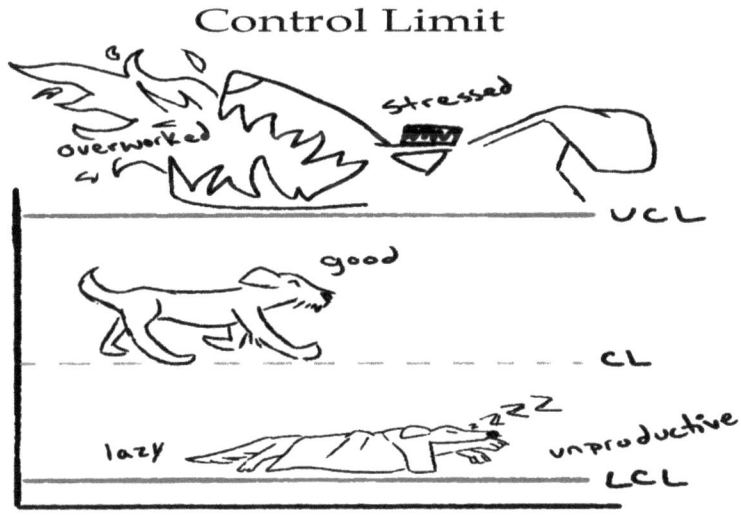

© The Author(s), under exclusive license to Springer Nature Switzerland AG 2026
E. Tokgoz, *Six Sigma for Continuous Improvement in Cybersecurity*, Synthesis Lectures on Engineering, Science, and Technology, https://doi.org/10.1007/978-3-031-91030-2_8

The control charts that we previously covered can help with the performance measurements of such systems. The "control" phrase can be understood in two different ways:

- **Controls in cybersecurity:** There are control mechanisms for keeping cyber security associated systems under control with the expected operational functionalities. Frameworks can enable an organization to manage security controls across different types of assets with consistency by using frameworks that are systems of controls [5].
- **Controls in six-sigma:** The Control stage after applying D-M-A-I steps is the finishing touch for the project: A system that doesn't maintain the improvements is a system that doesn't do any good for that system. This is the original meaning of "Control" in DMAIC. Hence, sustainability of the improvement that remains under control is a critical element of the six-sigma project. After improvements are made, statistically speaking, we don't want too much variation in the cybersecurity system. The variation in the system would cause the system to be away from its expected performance. The more varied, the more that we would face issues with the security of the system.

We start out by covering the control mechanisms that may or may not be the key point of a sustainable control phase for the six-sigma project. As for the examples, if we want to find the right level of control mechanism for an organization then it would require both (1) and (2) listed above. If we work on a six-sigma project that is more on the business side of the cybersecurity solution that doesn't require the use of (1) then (2) would have a standalone application.

8.1 Control Mechanisms in Cybersecurity

There are two key elements of control mechanisms in cybersecurity applications:

1. Essentials for setting up the control systems
2. Types of control mechanisms that can be employed for empowering the systems in place.

8.1 Control Mechanisms in Cybersecurity

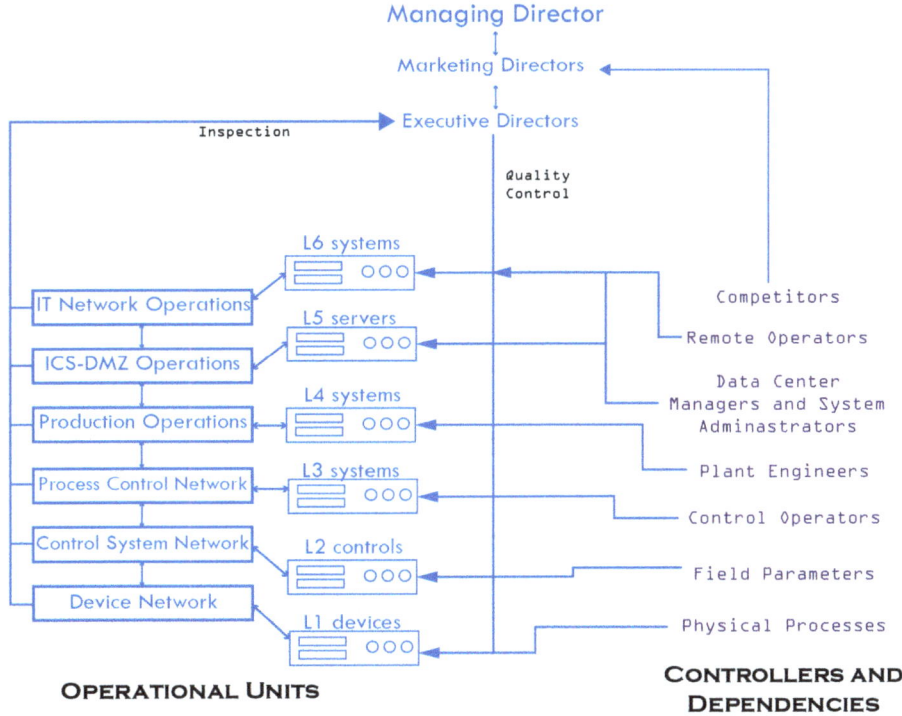

8.1.1 Essentials for Setting up the Control Systems

Some of the important considerations in setting up the control systems include the following:

1. **Entities:** Determining the existing sizes, capacities, performances, capabilities, weaknesses, strengths, durations of performance, and timings of the cybersecurity related systems' components. In this case we refer to personnel, machines, equipment, software, hardware (such as computers, servers), and cloud-related considerations as components. These determinations will help with the true control mechanism needs of the organization.
2. **Controls:** Determining the type and number cybersecurity challenges faced with the methods of control mechanisms utilized for declining these challenges is needed. These identifications will outline where your organization stands with the control mechanism.
3. **Sourcing:** An important aspect of reacting to something out of control is the source of control which can be internal or external. It would be a good strategy to identify the internal control resources with their performances and external sources with their performances. If there are resources that have overlapping reaction mechanisms to threads,

then the comparison of these resources will help you identify waste. These considerations would also depend on how tidy your organization wants to keep responding to threads.

4. **Accessibility rights:** Another critical design component of cybersecurity control mechanisms is the given rights for access. For instance, if a university president has high level of access rights to all the university functions, then hacking the president's account would open a door to many threads to many different functions. The control mechanisms may vary depending on the access rights given. These accesses can be categorized according to the information systems available and the control mechanism essential for such an information system. The optimal security control levels (i. e. ideal control level) depending on the rights given to personnel (such as very low, low, medium, high, and very high) need to be determined, these rights impact day-to-day operations of the personnel. The accessibility rights not only impact the organizational costs but also impact on the confidentiality, integrity, and availability of the organization along with the stress levels of the personnel. It also has an impact on the company's marketplace reputation.

5. **Budgeted?** While cybersecurity professionals might be solely focusing on the thread responses and handling instances, keeping the corresponding budgeting of them is also essential in a project.

6. **Controlled?** Determining the right control level for organizational excellence can be a black belt six sigma project itself. It would require a strategic data collection method and analysis of the data. Such a six-sigma project may require a business analyst, cyber security professional, and upper management personnel.

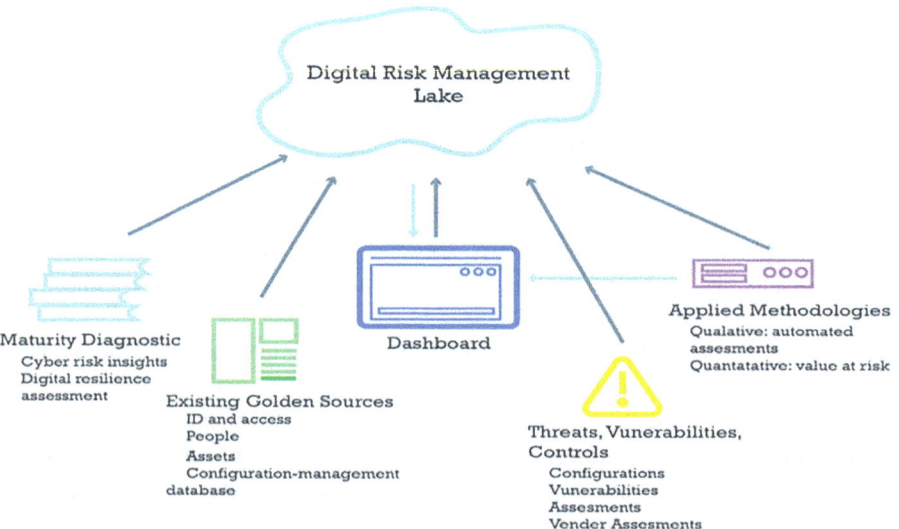

8.1.2 Organizational Control Mechanisms

The control types that we will focus on in this section include some of the considerations we have had so far; however, we will be adding more to these for establishing and reacting to out-of-control conditions.

8.1.2.1 Essential Employee Awareness and Training

It all starts with the employees of the company. The training of employees at an essential level is probably the most critical component for keeping cyber security systems under control. This is the sustainability for keeping the "human factor" on such systems under control through reminder mechanisms. The training is expected to increase employees' awareness of company-related issues and knowledge of the new issues that might be arising within the company. The right periodicity of the training is essential because it can be overwhelming for the employee otherwise. These trainings are expected to help with detection, responding, and recovering to the cyber threads. Control mechanisms that need to be emphasized should include.

- How to create a good password.
- Organization approved software programs to be used by legitimate vendors.
- Data and password storage policies.
- Data and password sharing policies.
- Appropriate company network use for external resource access.
- Malicious link and attachment recognition contained in spear-phishing emails.
- Control over social media is used to prevent from attacks executed through angler phishing attacks.
- Mobile and computer app download rights and permissions.

8.1.2.2 Healthy Information Technology Structure

It is the next important level of establishing the right control mechanism: the right IT structure. It is essential in many businesses for the development of IT with knowledgeable personnel who can maintain control over the cyber security related issues under control. An increase in the number of instances needs to be kept under control by IT. It is also critical to train IT to advance their knowledge in threads and expected responses to maintain a healthy business environment.

8.1.2.3 Right Combination of Software Usage

The essential responses to the cyber threads would certainly be initiated by the software purchased by organizations. As we pointed out previously, training of employees and the healthy IT infrastructure both play an important role in utilization of the software used for cyber threads. If the software is not chosen right (which could be a result of IT choices) then the training of employees can be absolute because the vulnerability of the systems to threads will be high. If too much security is placed with too many software packages, then the cost to the company increases (which causes waste) along with the redundant and overcomplicated software rights to access systems and much more complicated network infrastructure. The choice of right software packages with the needs determined as a part of six-sigma applications would yield to an optimized software package use. Application of antivirus solutions at a base level is essential in the software used. Updating these antivirus programs against different families of malware, including spyware, ransomware, worms, and trojan horses are also important for ever changing structures of the attacks. The creation of a threat database for tracking cyber-attacks that an organization face is the first level of awareness for the threads faced by the organization.

8.1.2.4 Automated and Manual Controls for Incidence Responses

Detection of threads through automated programs/software allows one of the best mechanisms for controlling and reacting to threads. In some cases, the detected cases require manual attention, and this attention is essential for the personnel to act upon. If the right actions are not taken on time, then the control mechanism can be redundant since there will be breach in the system. As we pointed out previously, the right combination of automated and manual control mechanisms to act is essential. This can also be a six-sigma project in which case the defined problem can be improvement of automated and manual controls in place for the organization.

For instance, security information management systems actively monitor, detect, and respond to security threats. Use of such systems allow security teams to keep track of all activities at the system or network level while security teams should assign responsibilities to security teams at the right level based on team members strengths to have control over incidence responses.

8.1.2.5 Dynamic Controls

Cybercrimes are advancing with the utilization of ever-growing use of technology including artificial intelligence and automated visual detection mechanisms. Responding to these threads requires active planning of incidence response depending on thread detection and prevention. What we so call "dynamic controls" are the controlling mechanisms that require adoption to the detected threads throughout the system analysis.

8.1.2.6 Perimeter Defense Controls

Cyber-attacks on the network through the internet are prevented by perimeter defense mechanisms. There are several perimeter defense controls utilized by organizations [4]:

- The use of firewall is the conventional control mechanism as the perimeter defense, and it can be a combination of both hardware and software. Its main role is to detect and react to suspicious traffic instances that are interacting with the user's internet utilization.

- A control mechanism on the perimeter defense can control the Wi-Fi connections. Organizations often provide employees and customers with public Wi-Fi access [1] which is insecure. Controlling mechanisms over the use of Wi-Fi needs to be established for secure access to the system.
- Domain Name System (DNS) is a control mechanism that prevents unauthorized web access to the organization's network. This control system allows the security of devices along with firewall protection. Filtration and restriction on content to be displayed are controlled by using DNS.
- Special access methods such as virtual private network (VPN) are a structured control mechanism for preventing breach through unsecured external networks. It is a professional way to establish a secure connection between the organization and any other network that may be insecure.
- An automatic control mechanism for protecting users' credit card information from security breaches is provided by Payment Card Industry Data Security Standards. It is possible to separate the terminals used for such transactions from the public and private networks [1].
- Automation of threat detection for Intrusion detection systems by using optimization and artificial intelligence to warn system administrators [8].

8.1.2.7 Control Over System Level Password Choices

Control over a system-level password choice might have to be enforced by the organization noting that the same use of password with simple password choices increases vulnerability of the system. This control mechanism can be designed in a way to overcome repetition of the same password for various locations and subsystems. These choices also need to be optimized for establishing the right control mechanism.

The ordinary settings of information technology products can cause password associated vulnerability. The repetitive nature of default configurations is very well known and therefore can cause weakness of the system for cyber-attacks. The best way to prevent this vulnerability would be differentiating the local mechanisms' infrastructure from the ordinary. Improvement of this repetitive behavior can also be a six-sigma project in a way to improve the design of the network. Many organizations today either use two-factor or multi-factor authentication for internal system access or require password choices that have certain length and a combination of characters, numbers, and letters with an additional pin entry required (e.g. a code sent to cell phone or Authenticator based access). The required periodic change of password is also another level of controlled protection for the system setup by system administrators.

8.1.2.8 Controlling Data Storage

Data storage can be in a local device, cloud based, or external hard drive driven. All these data storage methods (portable and non-portable) would require different control mechanisms for accessing the corresponding database. Some organizations only allow their employees to use their devices for work purposes and nothing else to minimize the risk. Use of cloud storage requires additional cautions and control mechanisms to be placed. Password protection and securing the hard drive locations for theft also play a role in protection of data. Control over these requires many different levels of security clearances and password protection at different levels of employee involvement. Keeping track of the numbers of these resources, capacities, and storage power are essential to keep the data storage resources under control. Keeping these resources under control would help with minimizing inventory waste. Policies should be placed by the organization to dispose of the data storage resources.

8.1.2.9 Back-Up Control

Backing up the data is an essential process step for an organization to continue healthy operations. A healthy back up requires certain considerations where we can apply 5W + How technique we covered in the Analyze section:

- **When does** the data need to be backed up? The right time to back up the data is essential.
- **Where** should the data be backed up? A secure location is necessary to back up the data.
- **Why** would the data be backed up? The critical importance of the need for backing up the data should be known.
- **What** needs to be backed up? In a group of files, certain files might be essential to back up and some others may not be essential.
- **Who** should backup the file? An organizational methodology might have to be followed because if there is more than one person who could back up the data and if the person who is supposed to back up the data is not known then this could cause organizational problems. It could also overwhelm the employees if the same person does it all the time in a group of staff with the same responsibilities.
- **How** frequently does the data change that requires back up?—Knowing the frequency of data backup helps with reserving time and space recourse.

The roles should be clearly defined to have a health, professional, and organizational back up plan. As simple as it may sound, it is important to organize this work noting that the files are expected to be backed up throughout the years. File backup is essential in case of a security breach.

In the case of cloud technology use, a practical choice for storing backup data is provided. The choice of strong passwords becomes more critical in this choice to secure cloud backups and other access control measures.

8.1.2.10 Controlling Risk

Risk assessment based on the use of metrics also needs to be controlled over time. Control charts can be particularly beneficial for assessing and establishing such a measurable control mechanism that is driven by time. Several risk metrics can be designed to understand whether the risk metrics success in company operations and cyber thread responses.

Example Robust cyber-security techniques are necessary to apply in modern industrial control systems and other cyber-physical systems (CPS) such as smart grid, unmanned vehicles, manufacturing plants, chemical plants, and nuclear reactors are complex interconnected systems with extensive cyber and physical components. Computing, communications, and network based cyber-attacks make CPS vulnerable to process-aware attacks that aim to disrupt the proper functioning or obstruct performance, efficiency, stability, and safety of the CPS. Process oriented attacks can impact the functioning of CPS thereby harming the performance or stability of the overall system or its components by modifying the information flow or the computational behavior of the system. Increasing network connectivity of the computational nodes of a CPS facilitates the maintenance and on-demand re-programmability of computational nodes. However, this increasing connectivity raises the potential for cyber-attacks that attempt unauthorized modifications of the run-time parameters or control logic in the computational nodes. One way to have control over the CPS operations is to develop effective real-time attack monitoring and threat mitigation mechanisms [3].

8.1.2.11 Controlling System Access

Controlling access is a way to control security level based on user authentication. The role and required system level access of an individual within an organization determines the system access rights of that individual. The variety of control measures in an organization is driven by the security concerns within the company along with the roles of the individuals.

For instance, the president of an organization is expected to have higher level access to any other member of the organization which is an example of a least-privilege access control that allows an organization to protect sensitive resources from unauthorized control. The distinguishment of privileged accesses needs to be also established by using the least privilege access approach so that vulnerability of the system is reduced. For instance, if A and B are two different departments with no overlapping access needs within the system then they need to have different access rights to the system. This mechanism allows the minimization of resource waste. Administrative accounts should only be given to system administrators so that employees don't interfere with administrative processes. Employees should have their own accounts and enforce password security options to achieve organizational transparency and accountability.

Another way that an organization may allow system level access is through mobile devices. Organizations can utilize Enterprise Mobility Management (EMM) systems through which enhanced business features can be accomplished with central mobile device management. EMM allows managing, auditing, and supporting the use of mobile devices [1]. There are mobile applications utilized by some organizations in which case source of download, near-field access (e. g. Bluetooth), allowing automated connectivity (i. e. saved password in a device) and access through public networks can cause cyber security issues within the system. Therefore, risk can be minimized (or controlled) by downloading cells phone applications from trusted resources, declining near-field access, eliminating automated access, and careful use of public networks. Policies need to be set up by the organization with the corresponding training provided.

8.2 Control Mechanisms in Six-Sigma

A system's operational performance can be measured by using control charts to understand whether there are "out-of-control" (i. e. not within the 6σ range) or not. These control charts, as we previously covered in the Analyze chapter, allow us to observe whether the system is under statistical control or not through numerical observations of average and standard deviation. In the case when there are data points that are not under 6σ control limits, the system indicates unexpected behavior. Improving the process is expected to eliminate these out-of-control points because of the changes made in the system. After the improvements are made, the data needs to be recollected and analyzed. The points that were observed to be out-of-control are now expected to be under control. If there are data points that are still not under control, then it is also important to determine under what conditions this situation occurs and the best way to signal the problem along with the determination of the solution. This is one of the reasons for six-sigma to be a continuous improvement; the system elements would almost always either have out-of-control points or too much variation in the process that require improvement. There are key elements of control section that one needs to keep in mind such as.

- Control Charts redesigned
- Reacting to Out-of-Control
- Documentation
- Finalization

After determining the out-of-control data points in the process by observing control charts, it is important to set up a signal to point out this situation. Documenting the findings of the Control phase and finalizing the project are the last two concepts to be covered in this section.

8.3 Designing Charts for Controlling

Systems are designed to respect certain tasks expected by the business to fulfill ordinary tasks. In certain cases, these systems have out-of-control behavior and such behavior that needs modification therefore a nature need for such systems is to determine limits to decide whether the system is under control or not. If you think about it, control limits are typically a part of our daily life. On occasions, we need to keep operations under control and there is an upper and lower control limit to these operations that need to be fulfilled. As the decision maker, you have the judgement of under control conditions based on your way of living. Therefore, it is not so strange to have control limits for systematic applications in cybersecurity, however such limits may typically be expected to be measurable/numerical. Two types of control charts that we will cover in this section are the following:

- Continuous
- Discrete

As we covered in the first chapter, the Greek letter σ which we write as "sigma" in English is used for representing the statistical standard deviation in mathematics. Hence, 6 sigma can be recognized as the six standard deviation range within an average value. The coverage of control charts in this section would utilize sigma ranges from the standard value that would also incorporate the data values to tell us whether the data is under control or not that is associated with sigma deviations from the standard.

8.3.1 Basic Definitions

There are basic terms that we need to define prior to starting the coverage of the control chart mathematically. We will use the following abbreviations throughout this chapter:

- **UCL.** Upper control limit
- **LCL.** Lower control limit

Some of the essential considerations in this chapter include the following:

1. The quality analysis conducted has a statistical nature because quantifications depend on statistical formulas such as mean, standard deviation, etc. of the data as well as the statistical distributions of the data due to uncertainty.
2. Statistical quality analysis of control charts typically relies on UCL, LCL, average, and standard deviation values.

3. Significance of statistical quality can be measured by Control charts and such analysis is helpful for visualization of the data.
4. Outliers in a control chart indicates "out of control" points of the process. Alerts can rely on such points.
5. Control limit documentation and its analysis helps with determining unusual behavior within a system and the triggers of such behavior.

Control helps to measure whether improvements worked or not after applying improvement while they can also be used for measuring the out-of-control points prior to improvement during the Analyze phase of DMAIC as well. The following are essential points for keeping in mind for quality control analysis during the data collection before and after improvements:

- **Data collection prior to Improvement Applications.** The data collected prior to application of improvements need to relate to the scope of the project and they should be fulfilling the objectives of the project.
- **Data collection after Improvement Applications.** The data collected after the completion of improvements require several steps to be taken to ensure that the project will be successful for measuring the effectiveness of the improvements from an operable perspective under expected control measures.

1. Data collection from the same sources is crucial in the way it was collected from the system; Such data recollection is necessary after improvements to determine the processes' statistical control.
2. Post-improvements rely on analysis of the recollected data as an indicator of how much the improvements impacted the processes in place.
3. Pre- and post-improvement stages' data need to be relying on specifications made for the system:
   a. System's pre-improvement phase data collection needs to be documented in detail with the methods of data collection applied. Answering 5W & how can help in this setting.
   c. The quality of the data needs to be high for reliable results.
   d. Pre- and post-improvement data sets need to rely on the same sources and follow the same sampling technique when possible. Deviation from the pre-improvement data collection technique for the post-improvement data collection (unless the changes require so) may result in attaining measurable results that are incompatible. In the case when the same data collection method's application is impossible, the data collection methodology that is closest to the one that was chosen at the pre-improvement data collection methodology should be selected for the post-improvement data collection. For example, if the pre-improvement data is collected

8.3 Designing Charts for Controlling

from a particular server per hour during certain hours, the same data can be collected from the same server every hour during the same time period during post-improvement analysis of the process.

4. Monitor the processes and collect one or more data sets that represent the improvement goals.
5. Systems may have specific periods with special operations and such periods may produce different results than the typical systematic behavior. Distinguishing such behavior from the ordinary behavior as such data points would cause outlier points within the data set. Close attention should be paid to such special cases that might impact on the quality of the data. For instance, a server may be under maintenance therefore its service may not be much lower than expected during that time frame.
6. The sample size needs to be a significant portion of the population for a true representation of the population.

The following basic terms and notation would be used during control charts' coverage:

- The overall data K is assumed to be formed by several data sets.
- We assume there are m data sets (called X_i) and these sets contain data points that do not overlap (i.e. independent) with the collection of these sets form the data set A:

$$A = \{X_1, X_2, \ldots, X_n\}$$

The number of data points in X_i may or may not be the same.

- $\overline{X_k}$: Average value of data values in set X_k
- $\overline{\overline{X}}$: Average value of $\overline{X_k}$:

$$\overline{\overline{X}} = \frac{1}{n}\left(\overline{X_1} + \overline{X_2} + \overline{X_3} + \cdots + \overline{X_n}\right)$$

Recall. The range R of a data set can be determined by calculating the difference between the maximum and minimum value within the data set:

$$R = \max(X) - \min(X)$$

with max(X) representing the maximum and min(X) is the minimum values in the data set.

- Now that we have several subsets in set A, we will modify R and use the following for calculating the averages of each subset

$$R_k = \max(X_k) - \min(X_k)$$

with

$$\max(X_k) : \text{Maximum value in } X_k$$

$$\max(X_k) : \text{Minimum value in } X_k$$

- The average value of all the ranges of subsets of A, namely $\overline{R}$, can be calculated by using the range values R_i:

$$\overline{R} = \frac{R_1 + R_2 + R_3 + \ldots + R_n}{n}$$

The next section initiates the coverage of control charts for continuous data.

8.3.2 Designing Control Charts for Continuous Variable Data

There are three control charts that can be designed for variables that have continuous nature:

- **$\overline{X}$-R Charts (Xbar-R charts).** There are two charts with one structured for the average of subsets and the other one structured for range values of the subsets. This type of chart can be typically used for subsets with sample sizes 4, 5, or 6, and it can be at most 10.
- **$\overline{X}$-S Charts (Xbar-S charts).** There are two charts that consist of two graphs with one structured for the average of subsets and the other one structured by using the standard deviation (s) values of the subsets. This chart can be used for quality control determination when sample size is greater than 10.
- **IMR. As a special case,** this chart is used when the sample size is 1.

There are certain precalculated values used for identification of control chart values for $\overline{X}$-R, $\overline{X}$-S, and IMR charts, and the Table 8.1 outlines such constant values up to 10 sample sizes.

8.3.2.1 $\overline{X}$-R Charts
There are two charts that make up the $\overline{X}$-R charts with two graphs that need to be designed for them:

8.3 Designing Charts for Controlling

Table 8.1 Constant table values useful for control chart design

Control charts' constant values								
	XBar—R charts				XBar—s charts			
	Chart for average	Chart for ranges (R)			Average constants	Standard deviation (s) constants		
Subgroup size	A_2	d_2	D_3	D_4	A_3	c_4	B_3	B_4
2	1.880	1.128	–	3.267	2.659	0.7979	–	3.267
3	1.023	1.693	–	2.574	1.954	0.8862	–	2.568
4	0.729	2.059	–	2.282	1.628	0.9213	–	2.266
5	0.577	2.326	–	2.114	1.427	0.9400	–	2.089
6	0.483	2.534	–	2.004	1.287	0.9515	0.030	1.970
7	0.419	2.704	0.076	1.924	1.182	0.9594	0.118	1.882
8	0.373	2.847	0.136	1.864	1.099	0.9650	0.185	1.815
9	0.337	2.970	0.184	1.816	1.032	0.9693	0.239	1.761
10	0.308	3.078	0.223	1.777	0.975	0.9727	0.284	1.716
15	0.223	3.472	0.347	1.653	0.789	0.9823	0.428	1.572
25	0.153	3.931	0.459	1.541	0.606	0.9896	0.565	1.435

1. **R-graph.** Each set consists of subsets (even if the set doesn't appear to have subsets, single elements can be considered as a subset on their own with n = 1) with a range of within each subset's elements. This provides multiple range values for subsets and the R-graph is the graph of such ranges $\overline{R_i}$ of subsets that also contains $\overline{R}$, UCL, and LCL as lines of control limit indicators. Using D_3 and D_4 constant values provided in Table 8.1 (and please visit other resources such as [6, 7] if additional values for larger data sizes would be needed) we can calculate the upper and lower control limits of the graph.

$$UCL_R = D_4 * \overline{R}$$

$$Range = \overline{R}$$

$$LCL_R = D_3 * \overline{R}$$

2. **$\overline{X}$ (Average) Data Graph.** Noting that datasets may be made up of subsets of the dataset, $\overline{X}$ requires the calculation of each subset and this average becomes a point to be placed on the $\overline{X}$ graph represented by the $\overline{X_i}$ values; Therefore, the graph of $\overline{\overline{X}}$

contains all of the average values of the subsets with $\overline{\overline{X}}$ representing the average of the entire data along with UCL and LCL representing the upper and control limits placed as horizontal straight lines as the control limits. The number of data points in each subset is the main driver of the coefficients to be used in the associated formulas with the constant value of A_2 determined by using the sample size of the subset given in Table 8.1:

$$UCL_{\overline{X}} = \overline{\overline{X}} + A_2 * \overline{R}$$

$$\text{Center Line for Xbar Chart} = \overline{\overline{X}}$$

$$LCL_{\overline{X}} = \overline{\overline{X}} - A_2 * \overline{R}$$

Example 8.1 *($\overline{X}$-R Charts)* Returning to Example 3.14 with the coverage of Incident Time to Close, the interest is to be able to identify out-of-control stages on the associated Analyze, Mitigate, Contain, and Post-mortem stages to be able to identify extreme cases. We only take a sample of the overall data for utilization however the entire data can be selected.

Suppose our focus is on subsets of different incident reporting sources where issues occurred for only Analyze stage of incident recovery with three data points collected for each in seconds:

Set	Source	Cause	Analyze
1	Operations	Security Configuration Error	19537
1	Operations	Security Configuration Error	26263
1	Operations	Security Configuration Error	56454
2	Incident Response Team	Inadequate risk analysis	45081
2	Incident Response Team	Inadequate risk analysis	8166
2	Incident Response Team	Inadequate risk analysis	2496
3	Intrusion Detection System	Threat Actor Target Individual	72144
3	Intrusion Detection System	Threat Actor Target Individual	86438
3	Intrusion Detection System	Threat Actor Target Individual	21849
4	Operations	Inadequate risk analysis	18966
4	Operations	Inadequate risk analysis	1072
4	Operations	Inadequate risk analysis	88906

This data contains 4 subsets with each subset consisting of 3 elements and it is a very small data set that may not be reliable for analysis purposes; therefore, the following calculations and information development is only for practical demonstration purposes.

8.3 Designing Charts for Controlling

- **$\bar{X}$ graph**

  a. Considering the Operations source with Secure Configuration Errors cause, $\bar{X}_1$ and R values can be calculated as follows:

  $$\bar{X}_1 = \frac{19537 + 26263 + 56454}{3} = 30751$$

  $$R_1 = 56454 - 19537 = 36917$$

- Considering the Incident Response Team source with Inadequate Risk Analysis cause, $\bar{X}_2$ and R values can be calculated as follows:

  $$\bar{X}_2 = \frac{45081 + 8166 + 2496}{3} = 18581$$

  $$R_2 = 45081 - 2496 = 42585$$

- Considering the Intrusion Detection System source with Threat Actor Target Individual cause, $\bar{X}_3$ and R values can be calculated as follows:

  $$\bar{X}_3 = \frac{72144 + 86438 + 21849}{3} = 60144$$

  $$R_3 = 86438 - 21849 = 64589$$

- Considering the Operations source with Secure Configuration Errors, $\bar{X}_4$. and R values can be calculated as follows:

  $$\bar{X}_4 = \frac{18966 + 1072 + 88906}{3} = 36315$$

  $$R_4 = 88906 - 1072 = 87834$$

The average value of the subsets' range and average of the averages are the following:

$$\bar{\bar{X}} = \frac{30751 + 18581 + 60144 + 36315}{4} = 37281$$

$$\bar{R} = \frac{36917 + 42585 + 64589 + 87834}{4} = 57981$$

The value of $A_2 = 1.023$ as the subset sample size is 3 that can be found in Table 8.1. Using the calculated values, we can calculate the upper limit, e, and lower limit values:

$$UCL_{\overline{X}} = \overline{\overline{X}} + A_2 * \overline{R} = 37281 + (1.023) * (57981.25) = 96595.82$$

$$\text{Center Line for Xbar C} = \overline{\overline{X}} = 37281$$

$$LCL_{\overline{X}} = \overline{\overline{X}} - A_2 * \overline{R} = 37281 + (1.023) * (57981.25) = -22033.8$$

The lower limit is a negative number however we don't have any negative values therefore we can select zero to be lower limit instead. The $\overline{X}$ graph corresponding to this example is the following:

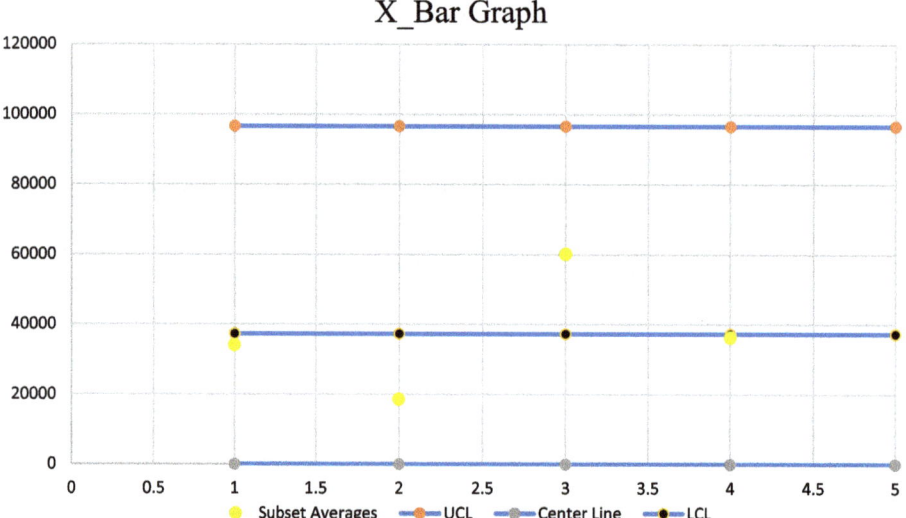

As is easy to observe from the above graph, the subset average points that make up the data set are within the UCL and LCL with values close to the center line (i.e. $\overline{\overline{X}}$) therefore there are no out-of-control points. We still need to note that the sample set size is small with a limited number of samples therefore the results may not be reliable. More data points are needed for more accurate results' attainment.

The R-graph values are the following by using Table 8.1 values for a sample subset size of 3:

$$UCL_R = D_4 * \overline{R} = (2.574)*(37281) = 149243.7$$

$$Range = \overline{R} = 57981.25$$

8.3 Designing Charts for Controlling

$$LCL_R = D_3 * \overline{R} = (0)*(37281) = 0$$

The graph representing the R-chart is shown below for this example. Just like the XBar graph, the range graph contains the subset range point values that make up the data set are within the UCL and LCL with values close to the center line therefore there are no out-of-control points. We still need to note that the sample set size is small with a limited number of samples therefore the results may not be reliable. In addition, the range values are too high within the data subsets that impact the results. More data points are needed for more accurate results' attainment.

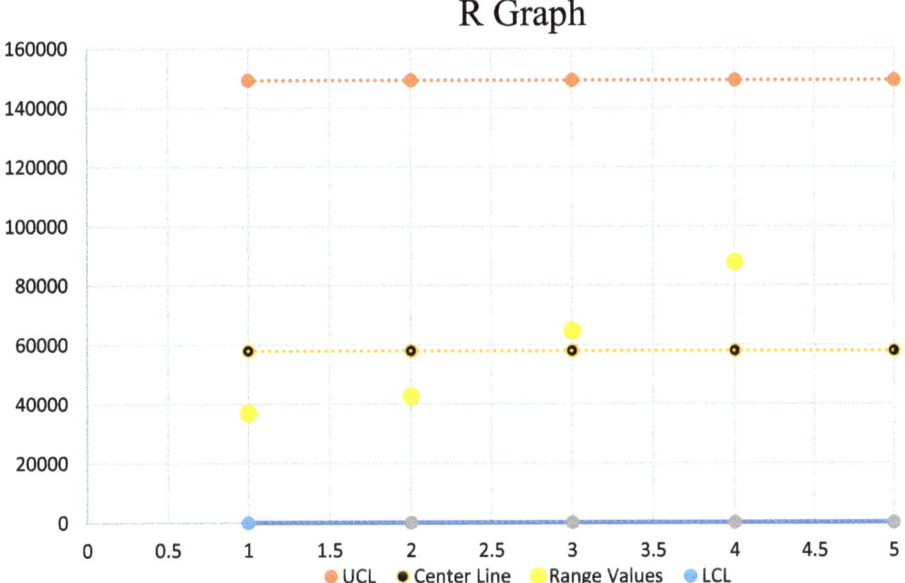

In these graphs, the meaning of "out of quality control" can be recognized as the values of the data points appear above the UCL. In general, there are two cases that can be considered as out-of-control conditions to occur:

- If UCL is lower than the values of the data points.
- If LCL is greater than the values of the data points.

Identification of out-of-control quality conditions help to assign causes to unexpected quality issues. There are no out-of-control points in the above given example.

8.3.2.2 $\overline{X}$-S Charts

In the case when we have a data set made up of samples collected from continuous data, s-chart can be used to analyze data set of sizes 11 or greater, or in the case when the data is represented by a variable. The standard deviation of each sample subset is represented by the letter s, and the average of these standard deviations is $\bar{s}$. The $\overline{X}$-s chart requires two graphs:

- **$\overline{X}$ graph.** This graph is similar to the one we constructed for $\overline{X} - R$ charts, except the $\overline{X}$ graph uses standard deviation values instead of the range values of the subsets in the UCL and LCL formulas with $\bar{s}$. representing the average of standard deviation of the subsets:

$$UCL = \overline{\overline{X}} + A_3 * \bar{s}$$

$$\text{Center Line} = \overline{\overline{X}}$$

$$LCL = \overline{\overline{X}} - A_3 * \bar{s}$$

- **s-graph.** This graph is an indicator of the subset's standard deviation changes with the corresponding UCL, center line, and LCL vu calculated by using Table 8.1 values of B_4 and B_3.

$$UCL = B_4 * \bar{s}$$

$$\text{Center Line} = \bar{s}$$

$$LCL = B_3 * \bar{s}$$

The following are examples of control charts attained by using the entire data set for the Analyze (on the left) and Mitigate (on the right) stages of the Incident Time to Close example. The top graphs are Xbar graphs while the bottom graphs are range graphs.

The individual value considerations and IMR graphs rely on single data point considerations that utilize different center lines as outlined in Table 8.2. This table contains a summary of the continuous variable control chart type formulas and the associated subgroup sample sizes.

8.3 Designing Charts for Controlling

Table 8.2 Formulas of control charts developed for continuous variables with the Σ sign representing terms' summation

Type of control charts	Subgroup samples	Center line	Control limits	
Average and range / Xbar-R	Constant and < 10, but usually 3 to 5	$\bar{\bar{x}} = \frac{(\bar{x}_1 + \bar{x}_2 + \ldots \bar{x}_k)}{k}$	$UCL_{\bar{x}} = \bar{\bar{x}} + A_2\bar{R}$	$LCL_{\bar{x}} = \bar{\bar{x}} - A_2\bar{R}$
		$\bar{R} = \frac{(R_1 + R_2 + \ldots R_k)}{k}$	$UCL_R = D_4\bar{R}$	$LCL_R = D_3\bar{R}$
Average and range / Xbar-S	Variable or ≥ 10	$\bar{\bar{x}} = \frac{(\bar{x}_1 + \bar{x}_2 + \ldots \bar{x}_k)}{k}$	$UCL_{\bar{x}} = \bar{\bar{x}} + A_3\bar{s}$	$LCL_{\bar{x}} = \bar{\bar{x}} - A_3\bar{s}$
		$\bar{s} = \frac{(s_1 + s_2 + \ldots s_k)}{k}$	$UCL_s = B_4\bar{s}$	$LCL_s = B_3\bar{s}$
Individual values and moving range / IMR	1	$\bar{x} = \frac{(x_1 + x_2 + \ldots x_k)}{k}$	$UCL_x = \bar{x} + \frac{3}{d_2}\bar{R}_m$	$LCL_x = \bar{x} - \frac{3}{d_2}\bar{R}_m$
		$\bar{R}_m = \frac{(R_1 + R_2 + \ldots R_{k-1})}{k-1}$ $\bar{R}_m = \|(x_{j+1} - x_j)\|$	$UCL_{Rm} = D_4\bar{R}_m$	$LCL_{Rm} = D_3\bar{R}_m$

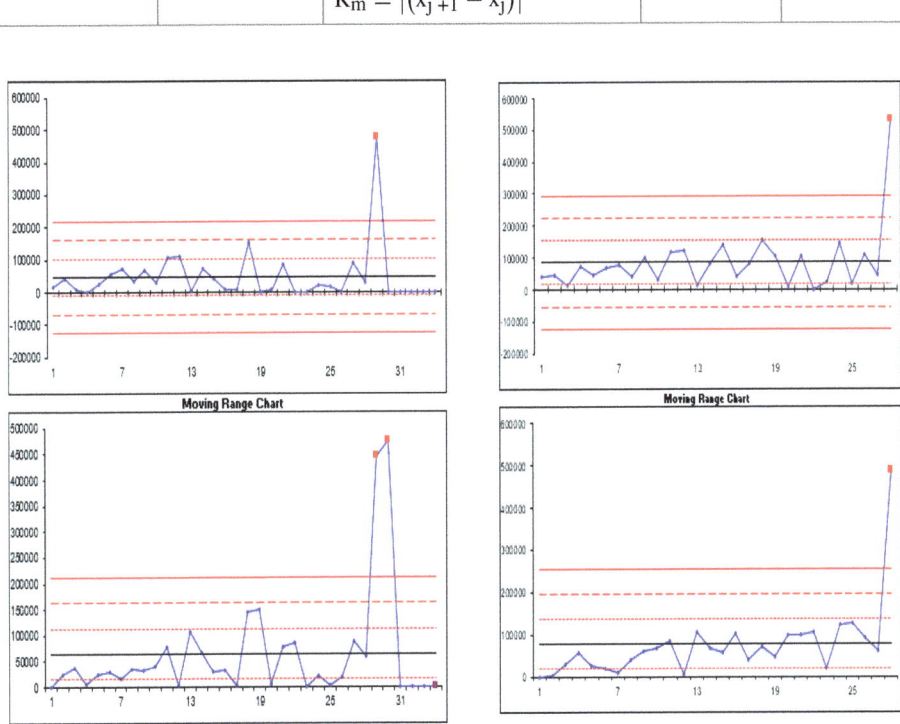

8.3.3 Control Charts for Discrete Data

There are two types of control charts used for discrete data:

- Defects per unit Control Charts (u and c charts)
- Defect Number Control Charts (p and np charts)

We can also classify such charts using the continuity of the sample size:

- Constant Sample Size (c and np charts)
- Continuous Sample Size (u and p charts)

Our focus in this section would be on these four chart types with the associated control limit formulas.

8.3.3.1 c Chart

Given a data et with constant subset data size we can use a c-chart for control chart's construction. This control chart can be used for representing the defective part numbers. To structure the control chart formulas, we use the following:

- d: Number of defects per unit in the sample
- q: Size of each sample subset (needs to be same for all subsets as a constant)

The control chart formulas for the Upper and Lower Control Limits as well as the Center Line can be calculated by using the formulas below for a c-chart. The $\bar{c}$. value represents the average non-conformities per subset. Noting that Six Sigma (6σ.) is identified by $+3$ standard deviations ($+3\sigma$) and -3 (-3σ) standard deviations from the Center Line, the data that falls within this range can be considered as "under control limit", and the other data points can be controlled as out-of-control data points from a non-conformity perspective with the assignable causes that should be investigated.

$$\bar{c} = \frac{\text{Total of all defects in the data}}{\text{Count of subset data}}$$

$$t_1 = \sqrt{\bar{c}}$$

$$LCL_c = \bar{c} - 3t_1$$

$$CenterLine = \bar{c} = \frac{d}{q}$$

$$UCL_c = \bar{c} + 3t_1$$

8.3 Designing Charts for Controlling

8.3.4 np Chart

Another control chart that we can design for constant sample size is a np-chart by using a size of more than 6. The control chart represents the number of nonconforming units. We use the following in the associated formulas:

- d_i: Number of defective parts in subgroup i
- k: Number of subsets
- n: Subset sample size

The control chart formulas for the Upper and Lower Control Limits as well as the Center Line can be calculated by using the formulas below for a np-chart. The $\bar{p}_i$ value represents the average non-conformities for subset i with i = 1,2...k. Noting that Six Sigma (6σ) is identified by +3 standard deviations ($+3\sigma$) and $-3(-3\sigma)$ standard deviations from the Center Line, the data that falls within this range can be considered as "under control limit", and the other data points can be controlled as out-of-control data points from a non-conformity perspective with the assignable causes that should be investigated.

$$\bar{p}_i = \frac{d_i}{n}$$

$$\bar{p} = \frac{\bar{p}_1 + \bar{p}_2 + \cdots + \bar{p}_k}{k}$$

$$t_2 = \sqrt{n * \bar{p} * (1 - \bar{p})}$$

$$LCL_{np} = n * \bar{p} - 3t_2$$

$$CenterLine = n * \bar{p}$$

$$UCL_{np} = n * \bar{p} + 3t_2$$

8.3.4.1 u Chart

It is natural for subsets of a large data set to differentiate and a u-chart can be used for such varying sample sized subset data for constructing control charts. This approach can be used for number of non-conformities per unit. Let

- a_i: Number of defects in the production of set i
- k: Number of subgroups
- q_i: Subset's (sample) size of set i

The control chart formulas for the UCL and LCL as well as the Center Line can be calculated by using the formulas below for a u-chart. The $\bar{u}_i$ value represents the average non-conformities per subset for subset i with i = 1,2...k. The UCL and LCL values in this setting are typically not straight lines and they look like a combination of broken pieces of lines' combination. Noting that Six Sigma (6σ) is identified by $+3$ standard deviations ($+3\sigma$) and -3 (-3σ) standard deviations from the Center Line, the data that falls within this range can be considered as "under control limit", and the other data points can be controlled as out-of-control data points from a non-conformity perspective with the assignable causes that should be investigated.

$$\bar{u}_i = \frac{a_i}{q_i}$$

$$t_3 = \sqrt{\frac{\bar{u}}{q_i}}$$

$$LCL_u = \bar{u} - 3t_3$$

$$\text{Center Line} = \bar{u} = \frac{\bar{u}_1 + \bar{u}_2 + \cdots + \bar{u}_k}{k}$$

$$UCL_u = \bar{u} + 3t_3$$

8.3.4.2 p Chart

Another type of control chart that we can construct for varying subset sizes is the p-chart for a subset size of 6 or more. Let.

- d_i: Number of defective parts in subgroup i
- k: Number of subgroups
- q_i: Sample size of subgroup i

The control chart formulas for the UCL and LCL as well as the Center Line can be calculated by using the formulas below for a u-chart. The $\bar{u}_i$ value represents the average non-conformities per subset for subset i with i = 1, 2...k. The UCL and LCL values in this setting are typically not straight lines and they look like a combination of broken pieces of lines' combination. Noting that Six Sigma (6σ) is identified by $+3$ standard deviations ($+3\sigma$) and -3 (-3σ) standard deviations from the Center Line, the data that falls within this range can be considered as "under control limit", and the other data points can be controlled as out-of-control data points from a non-conformity perspective with the assignable causes that should be investigated.

$$\bar{p}_i = \frac{d_i}{q_i}$$

8.3 Designing Charts for Controlling

$$t_4 = \sqrt{\frac{\bar{p} * (1 - \bar{p})}{q_i}}$$

$$LCL_p = \bar{p} - 3 * t_4$$

$$CenterLine = \bar{p} = \frac{\bar{p}_1 + \bar{p}_2 + \cdots + p_k}{k}$$

$$UCL_p = \bar{p} + 3 * t_4$$

As you have probably observed, the UCL and LCL formulas for the four control chart types we covered follow a conventional calculation based on the center line with $+3\sigma$ and -3σ deviations. This is because we are using the 6σ deviation from the average. If we want to also measure the 2σ and 4σ ranges, we can observe the $+\sigma$ and $-\sigma$ deviations and $+2\sigma$ and -2σ deviations from the average respectively.

A summary of the LCL, UCL, and Center Line formulas for p-, np-, u-, and c-chart are summarized in Table 8.3.

Table 8.3 Types of control charts with their sample sizes, center line, UCL, and LCL formulas where Σ represents the summation of terms

Type of control charts	Sample size	Center line	Control limits	
Proportion defective parts p-chart	Variable usually n > 5	$\bar{p} = \frac{\Sigma \hat{p}_i}{k}$ whereby $\hat{p}_i = \frac{\text{\# defective parts}}{n_i}$	$UCL_p = \bar{p} + 3\sqrt{\frac{\bar{p}(1-\bar{p})}{n_i}}$	$LCL_p = \bar{p} - 3\sqrt{\frac{\bar{p}(1-\bar{p})}{n_i}}$
Proportion defective parts np-chart	Constant usually n > 5	$\bar{p} = \frac{\Sigma \hat{p}_i}{k}$ whereby $\hat{p}_i = \frac{\text{\# defective parts}}{n}$	$UCL_{np} = n\bar{p} + 3\sqrt{n\bar{p}(1-\bar{p})}$	$LCL_{np} = n\bar{p} - 3\sqrt{n\bar{p}(1-\bar{p})}$
No. of defects per unit u-chart	Variable	$\bar{u} = \frac{\Sigma u_i}{k}$ where $u_i = \frac{\text{\# defects}}{n_i}$	$UCL_u = \bar{u} + 3\sqrt{\frac{\bar{u}}{n_i}}$	$LCL_u = \bar{u} - 3\sqrt{\frac{\bar{u}}{n_i}}$
No. of defects per unit c-chart	Constant	$\bar{c} = \frac{\text{\# defects}}{\text{\# units}}$	$UCL_c = \bar{c} + 3\sqrt{\bar{c}}$	$LCL_c = \bar{c} - 3\sqrt{\bar{c}}$

8.4 Reacting to Out-Of-Control

Now that we know how to measure and identify the states and conditions of a system that may cause out of control by using control limits through control charts, we can determine the reasons corresponding to these data points that cause the system to be out of expected control limits. Such assignable causes to these data points that do not reside within the Six Sigma range should be analyzed with the use of a cause-and-effect diagram; such a diagram would reveal assignable causes to the reasons for the system to be out of control. The reaction plan to such uncontrollable stages of the system is critical to the success of the Six Sigma project. Two of the ways to react to a mistake in a process are going to be explained next.

8.4.1 Operation Halting Mistakes

There can be critical mistakes that may require halting operations during production or service, and it is ideal to assign a warning system in the case when such problems occur. The degree of the mistake may require changing the way to halt the operation:

- In extreme cases, complete disconnection of the system components from each other may be necessary for the safety of the organization. In such cases, plans should be known by everyone, and they need to be practiced.
- In some cases, halting certain operations is needed in certain areas of the organization. In such cases, the causes and effects of halting such operations should be identified immediately. Unless the reasons are obvious, the cause-and-effect diagram would help to identify the assignable causes to the problem that occurred.

8.4.2 Warning Requiring Mistakes

In certain cases, due to its low harm, certain mistakes do not impact the system at a high level. It is enough to produce a warning with a signaling mechanism in place when such a mistake happens in the system if there are no reasons to halt the operations as a continuation of such issues. An automated system is typically desired (is possible) for catching such mistakes and warning the personnel in charge of the operations. For instance, some of the vulnerabilities caused by a person can be mistakes that may require only a warning for the employee to pay attention to not repeating the same mistake. Such warning systems in cybersecurity exist and it can be supported by additional cybersecurity professionals and placing them in the right places. Manual systems can also be in place and such a system can be directly utilized by the personnel. Assignable causes to the mistakes that occurred can again be determined by structuring a cause-and-effect diagram.

8.4.3 Documentation and Finalization

As the last phase of the DMAIC Six Sigma approach, Control charts and statistical quality measures are used, calculated and measured with the essential explanations of the works completed. After the completion of this phase, the documents necessary to finalize the project should be prepared that contains the works completed and how they are completed. The documents that we need to prepare for finalizing the projects include the following:

- The paperwork should be addressing the needs of the specific groups that have access to the system and operate the system. Such improvements need to be specific due to respecting the Least Privilege rule for accessing systems; This will allow right people to be knowledgeable about the right system. The documents prepared by the Six Sigma team can provide guidelines for the staff to follow certain directions or methods to implement improvements in the system. If possible and applicable, physical copies need to be printed and places in access points of the system (e.g. servers).
- There should be graphic representation of the data analyzed such as charts, figures, tables, etc. that display the current and expected future designs of the process for the (internal or external) stakeholders. Such information that is visually appealing can help others to easily follow the expected improvement changes that are necessary with the corresponding expected systematic behavior.
  - For instance, a pie chart that demonstrates a portion of the system before and after improvements will display the difference of improvements in the workplace with percentages.
- The designed quality control charts with their explanations should be included.
- The identified wastes and their places in the system with the corresponding categories to eliminate these wastes.
- The types of improvements applied to the system and the respective areas within the organization.
- Documentation of the control mechanisms to be employed for quality control.

Such documents should be incorporated into the final report of the project at the level of stakeholders' expectations. It is essential to prepare a final presentation that outlines the overall project steps, the process improvements, the expected actions to be taken in the system, differences between the pre- and post-improvement systems, and the expected benefits from making changes within the system. The diagrams, charts, etc. placed in the report should be also explained during the presentation. The outline of such a presentation can include but not limited to the following:

- Foundational information on the project with the associated systematic behavior outline.
- Definition of the Six Sigma Project's problem.

- Explanation of the pre-improvement systematic process.
- Measurements—Data collected and why such data is collected.
- Analyze Stage—Analysis applied on the measured data.
- Improvement—Methods used for improvement.
  - NVA identified with their locations and how much they are eliminated.
  - New system's design with explanations of the essential functions of the business units
  - Quantities identified through measurable improvements
  - Benefits of applying post-improvement results of the Six Sigma project.
- Control—Explanations on how the post-improvements can be sustained.
  - Explanation of when the post-improvement process may be out of statistical quality control.
  - Employee-expected actions for establishing sustainability of the post-improvement design.
- Final remarks

One ideal feedback from the stakeholders would be to collect survey data based on their opinions on the project results; although this sense-check should be implemented throughout the project to be able to make sure that the project is on track. This would not be too difficult if the Six Sigma team is members of the organization who actively work within the organization while it may be somehow more challenging if none of the Six Sigma team members is working for the organization. Two-sided communication that occurs between the stakeholders and the Six Sigma team is an essential element of the success of a Six Sigma project!

Chapter 8 Exercises

Exercise 8.1 Give an example of a cybersecurity application to design a system that would require halting the operation. What type of halting mechanism can be designed for such a system.

Exercise 8.2 Is there an existing automated cybersecurity system in real life that requires only a warning. What type of infrastructure is designed for such a system's performance.

Exercise 8.3 Give a real-life cybersecurity application (existing or the one you determine) with a discrete variable that utilizes control charts. Explain the nature of such a chart with its elements and what it can be used for.

Exercise 8.4 Give a real-life cybersecurity application (existing or the one you determine) with a continuous variable that utilizes control charts. Explain the nature of such a chart with its elements and what it can be used for.

Exercise 8.5 One of the challenges in real-life cybersecurity documentation is the volume of the information with the extreme number of pages or complex graphs produced. How can such data be simplified to produce documents, charts and graphs that are easier to follow. Can we implement this as a Six Sigma project? Please explain.

References

1. Ten Essential Cybersecurity Controls by George Mutune, CYBEREXPERTS, https://cyberexperts.com/cybersecurity-controls/
2. NIST Special Publication 800-53 (Rev. 4). https://nvd.nist.gov/800-53/Rev4
3. Khorrami, F., Krishnamurthy, P., & Karri, R. (2016). Cybersecurity for control systems: a process-aware perspective. *IEEE Design Test, 33*(5), 75–83. https://doi.org/10.1109/MDAT.2016.2594178
4. Top 10 most common types of cyber attacks—https://blog.netwrix.com/2018/05/15/top-10-most-common-types-of-cyber-attacks/
5. Razikin, K., & Soewito, B. (2022). Cybersecurity decision support model to designing information technology security system based on risk analysis and cybersecurity framework. *Egypt Inform J, 23*(3), 383–404.
6. Tokgöz, E. (2024). Quality & Lean Six Sigma Applications for Industrial Engineers. Springer Nature, Switzerland. https://link.springer.com/book/9783031557392
7. Tokgöz, E. (2024). Quality and Lean Six Sigma for Engineering Technicians, Synthesis Lectures on Engineering, Science, and Technology. Springer Cham 978-3-031-44033-5. https://link.springer.com/book/9783031440328
8. Tokgöz, E. (2025). Artificial bee colony optimization techniques' utilization for intrusion detection systems' analysis. In: *4th IEEE International Conference on AI in Cybersecurity (ICAIC) proceedings.* https://ieeexplore.ieee.org/stamp/stamp.jsp?tp=&arnumber=10848880

Index

A
American Society for Quality (ASQ), 207
Analysis of Variance (ANOVA), 130, 211, 219, 225
Analyze, 167–225

B
Bar chart, 144, 145, 189, 194
Batch means method, 170–172
Bias, 121, 122
Binomial distribution, 45–47
Box plot, 147, 224
Business process mapping, 85, 86
Business requirements, 125

C
C, 156
C++, 156, 157
Cause and effect, 146, 195–196
C-chart, 278
Chi-square test, 135
Confidence interval, 135, 137
Constrained optimization, 172
Continuous distribution, 35, 36, 49–58
Control, 257–282
Control charts, 276–279
Control mechanisms, 259, 260
Correlation, 133, 167–177
Correlation strength, 178
Critical to measurement, 113
Critical to process, 114

Critical to Quality (CTQ), 2, 68, 70, 91, 92, 115, 116, 152
Critical to satisfaction, 116–118
CTQ defective parts in a million, 152, 154
CTQ hierarchy tree, 91, 92
Customer, 73
Cybersecurity process, 77–83
Cybersecurity requirements, 124, 125
Cybersecurity risk framework, 7, 10
Cyber solution design, 237
Cyber waste, 24–31
Cyber waste categories, 24–31

D
Data analytics, 231, 243
Data collection, 107–110
Defects identification per unit, 152
Defects in a million opportunity, 152
Define, 61–96
Design of experiment, 179
Discrete distribution, 35, 39–49
Discrete normal distribution, 43–46
Documentation, 81, 162, 168, 211, 237–239, 249, 250, 266, 283–284
Document design, 238

E
Effectiveness of improvements, 247
Employee training, 240
End goal analysis, 180–181
Equal variance testing, 182

Essentials of control, 259
Exponential distribution, 55–58

F
Failure mode & effect analysis, 183–184
Finalization, 65, 283–284
5S, 231, 233–235
5W and How, 168–170
Flow diagram, 83–85
Functional flow diagram, 86, 87

G
Geometric distribution, 47

H
Heijunka, 193
Histogram, 144–146
Hypergeometric distribution, 48
Hypothesis testing, 33, 129–131, 143, 219

I
IBM SPSS, 155, 157
Improve, 229–254
Improving design, 237–241
Improving duration, 231
Improving network, 236
Incident time to close, 56, 62, 169, 170, 174–176, 179, 182, 198, 272, 276
Input, 75
Input-Process-Output (IPO), 67–69
Interaction table, 184–187
Interface analysis, 187

J
Java, 157
Jidoka, 194
Just in time, 195

K
Kaizen event, 16–22
Kanban, 94
Kano model, 95, 96
Key performance indicators, 110–113

L
Lean methodology, 1, 3, 74, 193
Lean six sigma, 73
Lean techniques, 193
Linearity, 119
Lower control limit, 267

M
Machine learning, 243
MATLAB, 158
Measure, 99–163
Measurement matrix, 124
Measurement system analysis, 118–122
Metric requirements, 126–128
Microsoft excel, 155–158
Microsoft visio, 161
Minitab, 158
Mistake prevention, 250

N
Negative correlation, 173, 175
Neutral correlation, 173–174
NIST CSF, 7–10
Non-Value Added Activity (NVA), 2–5
Normal distribution, 34, 39, 43, 50–54, 129–130, 136, 137, 143, 223
Np-chart, 278, 279
Null hypothesis, 138–143
Number of operators, 149

O
Organizational control mechanisms, 261–265
Out-of-control, 278
Output, 76
Overall methodology effectiveness, 188
Overall system effectiveness, 150–152

P
Pareto chart, 189–191
P-chart, 278, 280
Perl, 220
Poisson distribution, 46
Poka yoke, 193, 241
Positive correlation, 173
Probability, 26, 34–58

Problem space, 72, 73
Process, 75, 76
Process efficiency analysis, 191
Project charter, 66, 87, 88
Python, 161

R
R and R-studio, 162, 163
Reacting to out-of-control, 261, 266, 282
Regression analysis, 204
Repeatability, 122, 123
Reproducibility, 119, 122
Resource analysis, 196
Root-cause analysis, 196–199

S
Sample size choice, 138, 141, 142, 269, 270, 276
Sampling techniques, 102–106
SAS and SAS studio, 163, 164
Security tool coverage, 58, 63, 64, 91, 146, 181, 189, 201, 202, 204
Simulation, 241, 242
SIPOC diagram, 70–72
Six sigma project, 21–24
Software, 155–164
Spaghetti diagram, 88–91
Stability, 119, 120
Statistics, 33–58, 211, 212
Strategy, 5, 6, 9, 63, 76, 106, 117, 179, 245
Suppliers, 77
Surveying, 102–106
SWOT analysis, 199–201

System capability analysis, 187

T
Takt time, 148, 149
Time analysis, 201
Time-driven improvements, 244–246
Transportation design, 237, 238
Type 1 error, 140–142
Type 2 error, 141–142

U
U-chart, 279
Uniform distribution, 53–56
Upper control limit, 267

V
Value Added Activity (VA), 2–5
Value analysis, 203
Value Stream Map (VSM), 92–94, 148, 193, 203, 250
Visual Basic for Applications (VBA), 214–216
VSM icons, 92–94
Vulnerability patching, 65, 84, 138, 145, 169, 175, 186, 187, 192, 194, 201

W
Warning requiring mistakes, 282

X
XBar-R chart, 270, 275
XBar-s chart, 270, 275–277

GPSR Compliance

The European Union's (EU) General Product Safety Regulation (GPSR) is a set of rules that requires consumer products to be safe and our obligations to ensure this.

If you have any concerns about our products, you can contact us on

ProductSafety@springernature.com

In case Publisher is established outside the EU, the EU authorized representative is:

Springer Nature Customer Service Center GmbH
Europaplatz 3
69115 Heidelberg, Germany

www.ingramcontent.com/pod-product-compliance
Ingram Content Group UK Ltd.
Pitfield, Milton Keynes, MK11 3LW, UK
UKHW050046210625
459921UK00003B/13